Food Industry and the Internet

Food Industry and the Internet

Drew Smith

Blackwell
Science

Editorial Offices:
Osney Mead, Oxford OX2 0EL
25 John Street, London WC1N 2BS
23 Ainslie Place, Edinburgh EH3 6AJ
350 Main Street, Malden
MA 02148 5018, USA
54 University Street, Carlton
Victoria 3053, Australia
10, rue Casimir Delavigne
75006 Paris, France

Other Editorial Offices:

Blackwell Wissenschafts-Verlag GmbH
Kurfürstendamm 57
10707 Berlin, Germany

Blackwell Science KK
MG Kodenmacho Building
7–10 Kodenmacho Nihombashi
Chuo-ku, Tokyo 104, Japan

Iowa State University Press
A Blackwell Science Company
2121 S. State Avenue
Ames, Iowa 50014-8300, USA

First published 2001

Set in 10/13 Berkeley Book
by DP Photosetting, Aylesbury, Bucks
Printed and bound in Great Britain by
MPG Books Ltd, Bodmin, Cornwall

DISTRIBUTORS

Marston Book Services Ltd
PO Box 269
Abingdon
Oxon OX14 4YN
(*Orders:* Tel: 01235 465500
Fax: 01235 465555)

USA and Canada
Iowa State University Press
A Blackwell Science Company
2121 S. State Avenue
Ames, Iowa 50014-8300
(*Orders:* Tel: 800-862-6657
Fax: 515-292-3348
Web: www.isupress.com
email: orders@isupress.com

Australia
Blackwell Science Pty Ltd
54 University Street
Carlton, Victoria 3053
(*Orders:* Tel: 03 9347 0300
Fax: 03 9347 5001)

A catalogue record for this title is available from the British Library

ISBN 0-632-05753-X

Library of Congress
Cataloging-in-Publication Data
is available

For further information on
Blackwell Science, visit our website:
www.blackwell-science.com

Preface

This is e-intelligence, not just for technical people, but for anyone who wants to get up to speed with the e-commerce revolution and see how it is done.

It is independent, accurate and covers the revolution up to the time of writing which is March 2001. A year ago this week, perhaps judgementally, Alta Vista finally announced that it would break the telephone monopoly and give free Net access, Unilever announced 250,000 job cuts because it was going to develop on line procurement at a cost of 200m Euros. A week in which a simple idea like lastminute.com leveraged hundreds of millions of pounds in funding. A year on, the Nasdaq has crashed, BT's monopoly is bending, other companies have followed Unilever's example and lastminute.com does not figure in the top ten most visited sites.

I have resisted the temptation to go back over the stories that have been posted in the last 18 months and amend and edit them. Divided into several sections (pioneers and turning points, stop press, analysis, e-commerce, the big players, USA models, net tricks and resources), the stories are in chronological order within each section because that is the story of the Internet. And if at some stages they err on the side of innocence or unknowing, then that too is a part of the story. They are news stories, written and compiled from day to day. Understanding how far any one of us happens to be along the knowledge curve about e-commerce is one of the great skills in itself. So it is only fair that you have a chance to make that judgement on us too.

When we began working on this object in May 1999 (it was originally a project to cover trends in the food industry until we realised that the biggest trend of all was already starting to happen), we presumed there would be hundreds of companies already trading online. Indeed I had another commission to write a book on the 1000 best food sites. But in truth that is not what we found.

I could find 1000 home spun recipe sites. Cooking seems to have been one of the early forerunners of Net activity. Indeed I might even be able to find 1000 recipes for watercress. But I was looking for rather more than that. Discounting business card-style elementary web sites which hardly bear the name e-commerce, there were surprisingly few active players, perhaps less than 100. The giant supermarket Safeway was just in the process of putting up its first corporate site.

It remains true even now that the e-commerce revolution is still in the melting pot. Large numbers of early web sites have been abandoned and left unmanned in hyperspace. Some, especially the information sites, are nonetheless still useful. Others are just obituaries to what might have been.

The other lesson which is both frightening and challenging is that the Net changes. It is not a level playing field.

Eva Pascoe, writing in the Independent, a quality British newspaper, noted the lack of innovation and creativity that now characterise what were once vibrantly dynamic Internet pioneers such as AOL and Excite since they have acquired the middle-aged spread of conventional management.

A similar charge might also be laid at the door of the venture capital houses as they peruse each new would-be investment opportunity.

The winners on the Internet will need funding, but they need it to be empowered, not straight jacketed. No amount of money – as conventional publishers have discovered – is any secure route to riches. That in itself, is part of the medium's fundamental allure and dynamic. Also that it is global, which does not just mean a world-wide market, it means, or will do, world-wide competition in ideas. Also, the Net so far has been customer-driven . . .

Drew Smith
Foxley
Wilts

March 6, 2001

Editorial Introduction

This is a story of lust and greed, ambition and vision, hope and the hopeless. Sex – yes we all know about that part of it. Money, lots of it. Power too. The chance to own the information superhighway.

And also a certain charming innocence about the way of the world. While the financial pages were agog with the floatation of lastminute.com, another story noted that Rupert Murdoch was re-organising his business interests and bringing his satellite TV companies in line with his Internet projects. One wonders just how much Rupert Murdoch might have been thinking of investing in lastminute.com or if he might, just maybe, not think there are better things to do with his money?

And while venture capital poured another £7 million into farmingonline.com in Europe, it is interesting to note that a major USA rival is owned not by a media company but by Monsanto, the Bio agricultural conglomerate that developed GM foods, themselves subject of vicious harrying on the Internet by pressure groups in America.

The figures are almost self fulfilling. A third of the $4.4 billion in venture capital going to European Internet companies in 1999 went to British startups. Some 700,000 names were registered. Also compelling is the evidence that Britain has stolen a competitive advantage on Europe. Britain spent 3.5% of gross domestic product last year on information technology, compared with 2.5% in France and 2.6% in Germany. And while Sweden is more wired – it spent 4.5% of GDP on information technology – its small economy doesn't have the heft of Britain's.

Another reason is simply money. London is becoming the main channel for financing the European Internet. British Internet outfits have received $342 million in private equity investments from financial institutions, more than 20 times the amount invested during the same period in 1998. That compares with $243 million in France, $125 million in Germany, and $20 million in Sweden.

London has another competitive edge in that it has traditionally had a grasp of new media be that TV, advertising, film or music, all of which the Net can just roll into one package, and there is plenty to suggest that the Internet will be a fertile playground. Especially when broad band makes its long-awaited arrival.

As the Net economy unfolds it will eclipse huge areas of the conventional economy in ways that are only just becoming clear and in some cases in ways which none of us, or only a very few hopefuls, know about as yet.

The value in a brand name can be stood on its head in hyperspace. On the one hand, a brand like Starbucks can suddenly be the beneficiary of a Net strategy that sees millions of dollars pumped into it as part of a deal with an online home delivery newsagent. Other names, though, can mean nothing. What compelling reason can there be for anyone to surf and visit (and pay money) the Mars Bar or the Smint web site? These are high street brands precisely because they are a cheap, portable instant commodity which have little relevance to the Net economy. A Net brand in whatever form needs to make sales of £50 or more to justify the purchasing decision. Or in Starbucks' case, maybe they just got lucky and caught the mood of the moment because 90% of espresso customers were deemed to be Net users.

Brands will have to find other means of expression on the Net that go much further than simplistic web sites.

Of course the ordinary surfer – now estimated at one in five of the population – will not always be convinced of the value of the Net. Until they find a site that really delivers the Internet experience for them. And the most obvious and purest form of Internet interactivity is simply . . . shopping.

These new consumers may initially be loyal to the first portal that took them upstairs, but that loyalty may likely wear thin in time compared to the allure of a list of bookmarks that are personalised, focused and approved as they become more sophisticated.

Tesco has been widely criticised by so-called experts for its gung ho move in to the front of the market place, especially for its decision to use in store pickers rather than wait (as Sainsbury has done) for dedicated depots.

But another reading has it that Tesco has been shrewd. The Tesco brand is a portal site in itself and has potential to roll out into other countries much more easily than store based expansion. The unthinkable – which is always a good thing to think about with the Net – is that the Tesco portal might be worth more eventually than the stores themselves.

Waitrose, though, has probably been the most conventionally astute in its use of the Net with its Waitrose@work scheme which delivers direct to employees of companies with more than 300 terminals and was claiming to be in profit almost within months.

Whether big brands like these can adapt swiftly enough to change is probably the more critical question. Somerfield lost its management team in the week after acquiring the home delivery service Flannagans and now find themselves facing their ex-employees running Simplyorganic.net, an organic dedicated supermarket based in Covent Garden.

The impact of the Internet on the high street may well be advantageous, to the food community. Unlike travel, estate agents and banking, it seems unlikely that Net shopping will replace food buying; rather, probably, augment and galvanise it. All the evidence to date is that the Net shopper veers towards the mundane everyday items or the very niche specialist products, but still likes to shop conventionally too. It may well be that the current drift towards more cafe, restaurant and casual eating will be accentuated, and that specialist foods may well start to flourish again via the Net. Much more likely that foods will retain their part of a new leisure-oriented high street.

The drink market may feel this pressure more with the Net once again tipping the balance back in favour of the small specialist, direct from the vineyard company, which can side step the middle market place and pass on delivery savings and cost savings swiftly. But as with food this is probably more of a shift within an internal market rather than a wholesale collapse of trade that faces other bricks-and-mortar businesses.

For drink brands especially, the global nature of the Internet offers a new opportunity to develop untapped countries and continents by a singular global strategy.

The major seismic changes of the Internet revolution are much more likely to be felt inside larger corporations with the use of electronic information, administration and purchasing transforming the look and function of the conventional office.

The sales meeting of the future will surely be electronic, probably via video conferencing. The marketing strategy for the future will surely be informed by using the Net to target the key relevant individuals. Market analysis is already available at a hugely more sophisticated level via supermarket Intranets for those companies prepared to pay for it. And as is already happening in New York, the key currency of the future may no longer be set prices but barter and auction.

The strong players will be those companies that have the logistics to deliver the 'last mile'. And those with visionary Internet strategies which to date are either secret or do not properly exist. E-marketing is still in nappies.

And the question still remains as to whether the market place might suddenly lurch towards shopping mall-style sites that could damage supermarkets' market share or if the nature of the Net will always be to explore and seek out small and more individual products.

Problems

The biggest problem most companies face in generating an e-commerce strategy is writing the brief.

The conventional forum of a design company pitching for a project against a set or even flexible budget hardly allows the time for anyone to consider the implications and follow through.

In some cases companies have been fortunate in falling on dynamic and far-sighted authoring houses and have been able to develop swiftly. Too often the strategies have been rushed on both sides with the result that the web presence ends up looking like what it is – half a jigsaw that does not really fit.

Any authoring house can produce a web site. But for an e-commerce solution there is a requirement for someone who has the dual understanding of what can be achieved technically on the Net, how that might be applied in different environments and also a grasp of the conventional bricks-and-mortar business itself.

It may sound extraordinary but the number of companies who have paid for expensive web sites but do not have anyone monitoring the e-mail responses yet is remarkably high. Sodexho, for example.

So the net is not as advanced as perhaps we have all been led to believe. It will be, but as yet designing sites, the number of people qualified to create good environments, the sheer greed on both sides (of web authors asking for £25k for a Friday afternoon's work and of managing directors who still believe they can get on the Net for nothing) compounded by unhelpful banks and over-optimistic approaches to venture capital are creating a huge backlog which is fanned by all this talk of new Internet millionaires. The really bad news for conventional business is that all these factors can be bypassed by predatory self-funded operators.

Eventually, the Internet is going to make a lot more millions for conventional companies which can apply an e-commerce strategy than it is for innocent start-ups. But only if companies can see the global vision and get in ahead of competitors around the world.

The first impact on any company is simply the potential savings of using the Net to cut back on bills for conventional marketing, buying services like forms, phones, electricity, even legal services.

These in turn lead to more radical innovations like how to use the Net for marketing and procurement.

The big headlines may be made by software or Net savvy innovations, but the bread and butter opportunities are low risk and available now.

E-commerce is a management issue. For, in almost any sector, management that does not protect its own position by putting in place an e-commerce strategy faces the likelihood of its share price going down and, almost for certain, a virtual competitor arriving on the doorstep, quite possibly armed with a few million pounds of venture capital funding to fritter away on marketing.

Easily written perhaps, but the panic such a situation can induce inside an organisation can be equally critical and lead to delays, in-house feuding and general chaos. Corporate terror, even.

And to compound that terror, the next generation of intelligent software is just around the corner which may make all the current thinking obsolete.

One reason many smaller companies have been able to grasp the opportunity of the Net faster than big ones is simply that they do not have time to mess around with such strategic issues. They have nothing to lose and can develop an Internet presence relatively quickly and effectively and give it time to grow.

Conspicuously, some of the most successful small players have also had the time to learn the Net themselves and therefore short-cut many of the problems faced by larger groups trying to go forward in teams.

The Net is a very personal one-to-one medium and as such is not overly kind to attempts at team building sites.

The Net also has the ability to adjust a market. Suddenly perhaps, as with GM foods, a campaign is orchestrated across America that resists the accepted tide of change. Or, if a fishmonger in Cornwall begins to trade on the Net then how many potential new customers may come to him for savings and bypass traditional channels. Or else the tools for change suddenly appear unbidden on the Net, such as successful consumer sites like Coffeekid.com which might suddenly transform sales of one espresso machine against another, simply because Yahoo decided to make it the web site of the week.

This is the new world.

Welcome.

Contents

Preface v
Editorial Introduction vii
Introduction xxiii

Part 1: Pioneers and Turning Points 1
Part 2: Stop Press! 7
Part 3: Analysis 71
Part 4: E-Commerce 100
Part 5: The Big Players 153
Part 6: USA Models 172
Part 7: Net Tricks and Resources 195

PART 1: PIONEERS AND TURNING POINTS *1*
The first year of e commerce was 99–2000 but pioneers were developing software and elementary web sites from as early as 95–96 and putting them into the unknown void where for many, the expatriate audience was the first sign of trading life.

Top chef introduces revolutionary e-cooking service 1
AC Nielsen launches Web research package 2
Interview: rise of an organic banker 3
Pink dot goes PDQuick 4
Sales training goes on the Net 5
Pringles pioneers voice mail ads 5
That Sainsbury's depot . . . how it works 5

PART 2: STOP PRESS! *7*
This is the story of the first year of e commerce revealed as snapshots – as they happened – of the hopes, innovations, successes and failures of players big and small.

Beer sales show 300% growth 7
Pizza wars online 8
Hotel bookings service gets off to flying start 8
New site takes on Chateauonline 9
Ex-Burger King team look to supply links 9
Row over cut price Net rooms 10

Flying noodles get 20 visitors a month 10
Herb farm is growing 11
Wine portal looks to open trade account 11
Titan portal nightmare as names clash 12
Gourmets showered with gifts 12
Directors savage credit card bureaucracy 13
Farmers launch spuds online 14
Europ@web takes slice of wine portal 14
Gourmet site picks up IBM award 15
Rare breeds shop goes virtual 16
Those 25,000 Unilever jobs – the real story 16
Remote poultry breeder gets 70 visits a day 17
Spicing up the net 17
UK launch for wine auction site 18
Case study: espresso is hot (part one) 18
Case study: espresso is hot (part two) 19
Fishmonger delivers within 24 hours 21
Virgin snaps up orgasmic wines 22
Food from Britain launches corner shop for small producers 23
Another kite mark to validate e-business 23
Domino sells 6000 pizzas a month 24
¡Olé! Here comes Spain! 24
Chocolate orders: five to one now off the Net 25
Farmers undercut supermarket prices 25
Computer company wins the free phone race 26
Diet foods boom on the net 26
Phone charge rip-off continues 27
Net trade persuades DHL to float off online services 28
Whittards looks to brand alliances 28
Coffee Republic signs up with Net cafes 29
Cyber slimmers clean out the larder 29
And not to be left behind . . . 29
Jamie Oliver splashes out on his name 30
Marketing 30
Barclays launches e-business payment scheme 31
Banner advertising flops 31
One million new users sign online in April 32
Not so free after all 32
Le Franc re-brands with Henrietta Green 33
Whoops! *I Luv* delays chef's launch 33
Mail order now taking half its orders on the Net 34

And what wine would you suggest for turbot, monsieur? 34
Chocolate tops the sales in mall 35
Seattle coffee profits head into cyberspace 35
Survey slams online services 36
UK market matures 36
UK grocery market now bigger than the rest of Europe 37
Tesco tests Sunday trading laws 38
As if the web did not have enough recipes... 38
Priceline to take on Europe 39
Over 55s drink 15 bottles of wine a week 40
...Which is why Richard Branson moves in 40
Internet will usher in the euro 41
After the Net café, comes the Net pub 42
Tesco expands into America 42
Lastorders.com sups up £2.5 million investment 43
...and Eggs is looking for a £10 million benefit 43
And now its 10 million, official 44
Secure ordering triples orders 44
Freshfood.co.uk expands organic range to 5000 lines 45
Organic butchers put their beef on the Net 46
Gourmet gift market flourishes 47
Fish shop catches the big PR 47
Coffee site is percolating 48
Chicken coops go international 48
Expedia sets up Internet restaurants at airport 49
Hacker hoaxes Safeway 50
Online diet site gets fat on Nestle deal 50
Management buys out just-sites 51
Tesco to win half of e-grocery market says survey 51
Campbell Soups ties in with Talk City for recipe exchange 52
Real time figures on online shopping 52
Whoops! Not that price for WalMart! 54
One hour delivery wars 54
Electrical store trails kitchenware 54
Europe to overtake USA in 2 years 55
SimplyOrganic gets £1 million 56
The beerhunter is bagged 56
French make a swift exit from the crowded online wine market 57
Asda parent puts all its global procurement online 58
London delivery services collapse 58
Interview: Fledgling gourmet site takes wing 59

Waitrose cuts £35 million deal to win the last mile 60
Tesco pips Foodferry.com for top rating 61
Spice giant predicts Web will boost profits by five per cent 62
Iceland taking £600,000 a week online 63
Net diet is endorsed 63
Tea sales shoot up 450% on back of Java 64
Hic! It's a virtual bar 65
Hole! Mex goes Tex 65
Tesco says 35,000 grocery orders a week come off the Net 66
Now Nestle tries its hand at recipes 66
Foodoo folds 67
Del Monte sells fruit direct online 68
More than 7000 register turkey protest 68
Now Asda lays out its battle plan 69
Eggs benefit scrambled 69
FoodfromBritain.com gets a new look 69

PART 3: ANALYSIS *71*

Reading behind the riot of ecommerce headlines can be like trying to foretell the future in the tea leaves. We look in depth at the real trends that are going to fashion and shape the world of ebusiness.

Analysis: supermarket samurais or kamikaze 71
Packers plan organic box delivery via supermarkets 74
Analysis: Net GM campaign starts to bite 75
Those denial attacks – chapter and verse 76
Analysis: Has the bubble burst . . . 77
Analysis: the rise of an Internet success story 82
Analysis: the story of a one man start up 83
Analysis: the top and tail of a fruit B2B 85
B2B portal brings in farmers 91
Tradingproduce.com unveils field to Web technology 93
Analysis: Cooking on line 94
Interview: USA model flourishes quietly in the UK 96
Analysis: the truth about Internet fraud 98

PART 4: E-COMMERCE *100*

Business2Business has already begun to change the way we work. If email was the first step, then the really big changes are just around the corner as the Internet shapes up to take over traditional ways of working.

He is 30, married and well educated . . . but she is catching him up 100

UK Net usage surges ahead of France and Germany 101
Language slows up European growth 101
Online farming site goes for £7 million 103
Brake Brothers to invest £12 million 103
Macdonalds buys into Food.com 104
New B2B food portal gets £2.6 million venture-backing 104
Grocery market to hit £2.3 billion in next four years 105
Nestlé and Danone tie up for e-buying 105
500,000 credit card details go AWOL 105
Europeans like online shopping 106
Internet Trading Exchange for food industry 107
Euro e-commerce sales to top £105 billion by 2005 107
Europeans buck the trend 108
Food portal swallows up fish industry site 108
New B2B site for organic 110
It is not all rosy in the B2B area 110
Welcome to the era of intelligent software 111
It's business business business 112
Underlying trade in e-commerce is booming 112
Advertising set to kick in as revenue model 113
Produce B2B picks up $20 million new investment 113
World of Fruit claims 300 companies have signed up 114
Sainsbury to put $15 billion of global purchasing online 115
Webvan buys HomeGrocer for $1.2 billion 116
Delivery charges holding up e-commerce 117
UK leads Europe 117
Big Mac pays online 117
American giants merge 118
Bean auction trial saves 15% 118
Wal-Mart signs up Shopsmart 119
USA B2B portal sets up Korean venture 120
Big Mac sets up trading portal 120
China sets up online food exchange 121
Sweetener auction saves 6% 122
Efdex hits the rocks 122
Kraft buys into EthnicGrocer.com 123
About EthnicGrocer.com, Inc. 123
B2B trail shows spectacular savings 124
Electronic boxes on the doorstep 125
Interview: The man behind efoodmanager.com 125
Produce B2B site moves to seamless logistics 127

Hewlett Packard invests $6.1 million in Foodmarketplace.com 128
California set up big wine play 129
Profile: Transora, the birth of a giant 129
But Nestle and Danone bid to keep Transora out of Europe 130
Fish site lures the money with discount rates on transactions 131
Foodtrader.com deals $15 million in new funding 132
Irish meat company develops full traceability software 132
One billion litres of wine sold on the Net in the last four months 133
Wine portal buys into negotiative software 134
ecfood.com launches new suite of online tools for purchasing 135
Foodvision.com forms B2B alliance in restaurant kitchen sector 136
Now even sweets go for B2B 138
Now smart companies post their results on line 139
Wine gets groggy on too many portals 139
PlanetRice takes on funding 140
Thailand goes for vertical B2B site 141
Tesco hooks its check-out tills to the Net 142
Agribuys opens London office 143
All purpose B2B seeks to rationalize market supply 144
Americans move on the Spanish speaking market 145
Net opens up Chinese market 145
Agribuys takes in $32 million 146
Nestle ties up Thailand 147
Farmers combine to harvest online advertising 147
Asian B2B conducts global deals 148
Now recipes merge with coupons online 148
Globalnetexchange claims $300 million of trade a month 149
Sainsbury's ties up media play with Carlton TV 150
Big Mac creates its own B2B portal 151
Gofish adds news tools 151

PART 5: THE BIG PLAYERS *153*

The corporate vision and application of e commerce for large scale companies such as supermarkets and major manufacturers is a sector apart. The European market is the goal, the massive logistical implications the obstacle.

Anger as retailers pass on IT costs to suppliers 153
Hazlewood draws the line on e-commerce 154
7000 new e-jobs at Tesco 154
Supermarket brands lag behind 155
Monster portal edges closer to launch 156

Big two lashed in consumer test 156
Attack on online supermarket prices 157
Sainsbury appoints Egg head 157
Banks hushing up security attacks 158
Tesco is going to fall over say Americans 158
Tesco defends in-store picking 159
Iceland warms to the Net 159
M&S go all frilly 160
Tesco to split off Internet arm? 160
Sainsbury launches razzmatazz down the email 160
Tesco's Net profit 161
Dark horse headhunts logistics wizard 161
Online grocers lashed in consumer test 162
Safeway snaps up home delivery company 162
New Tesco goes into mortgages 163
Tesco looks to go overseas 164
Asda to launch TV and Web shopping at the end of the summer 164
Whoops! Safeway takes down the photos on its website... 166
Jet and Londis sign up to new Dropzone concept 166
Sainsbury ties up deal with Carlton 166
Milkman floats e-delivery 167
The rise and fries of company shopping 167
Safeway trial Palm 3 software 169
LM solutions snaps up tracking software 169
Express buys into M-box 170
Why Somerfield had to drop its e-commerce 170
The Web's cup runneth over 171
Coffee merchant snaps up mall 171

PART 6: USA MODELS *172*

America is the home of the Net and as quickly as the UK and the rest of the world, especially Japan, catches up, the first generation of major portals and web initiatives came from across the Atlantic.

Credit card fraud down to 2% 172
UK catches up on USA 172
The tills are ringing in USA 173
Food lags behind in USA 174
Net shoppers demonstrate brand loyalty 174
Webvan gets $120 million backing 175
Coffee is the froth on top for e-newsagent 176

Grocer pays $60 million for AOL link up 178
Chocolate site aims for $5 million 178
Tavalo secures $36 Million in funding 179
Online meat exchange draws record response 179
Cybercourier wars set to hit London 180
Big retail shakeout predicted 181
Farm wars take shape in US 182
American grocery market set for hypergrowth 182
Handy maps for the delivery man 183
Health police nail dirty restaurants on the N 183
Ranching portal looks to customise the news 184
Priceline moves on Europe 185
American pilot B2B audit system 186
Foodbuy.com consolidates USA food service domination 187
200 jobs go at Food.com 187
Gourmet sites merge 188
Watching the harvest online 188
Chicago raids the European larder 189
More retail casualties as the American market gets tough 189
Named and shamed 190
e-vineyard slips through the wine ban embargo 190
American restaurant site raises $42 million 191
Safety police organise Net sweep to track down bogus animal feed 192
New York restaurants get their own B2B site 193
Online deliveries rise 30% thanks to the storms 194

PART 7: NET TRICKS AND RESOURCES *195*

The most revolutionary part of the Net are sites dedicated to transforming everyday tasks. They get less publicity than the high profile consumer dot coms but as tools for business this sector is likely to have a more fundamental impact on trading than any other.

Online help for startups 195
Texas offers the idiot cure 195
Domain name pain 196
Now you can talk and see down the Net . . . for the cost of a local call 197
New gizmo reassures e-shoppers 197
Psst . . . wanna buy a second hand PC? 198
Knock knock . . . here comes the e-bailiff 198
Mind out for this new piece of red tape 198
Whoops! Is that your dotcom name for sale? 199

Interview with the Law: so who owns the website? 199
The true value of email lists 202
Now beenz links down into the packaging 202
Surprise Net link-up between Sainsbury and Oddbins 203

Index 205

Introduction

The first headline we ran on efoodnews.com was '*Tesco puts on 10,000 a week*'. This was followed by a backgrounder on the Waitrose@work scheme which delivered to offices with more than 300 computer terminals. Both companies in their ways were pathfinders. The third story was about how Sainsbury's held back allegedly wanting to follow the American example of dedicated depots which two years on is now seen to have been a mistake as Webvan and the other online grocers hit steep problems with that approach.

Asda had a technical problem of its own – the letters themselves happened also to stand for the American Sleep Disorders Association, and sure enough that was what came up on all the search engines. Safeway was still busy getting its first corporate site together, let alone entering e-commerce. So was Harrods.

Among the smaller players in the UK, Thoby Young had transformed a small monthly mail order fish delivery service into a full blown organic shop on the Net at www.freshfood.co.uk. It was, he said at the time, the moment he started to sell potatoes by email that his business began to take off. The name organic had already been filched – not by a food company but by a design company. RoomService.co.uk was claiming the first online drinks licence so it could sell wine as well as restaurant meals delivered to homes and offices. Mike Jarman, who quite possibly set up the first UK food site back in 1995 at www.botham.co.uk suddenly found that Americans were arriving at his Whitby Café to introduce themselves having found him through his web site and decided to visit.

There were other Net tricks too – mobile phones were 25% cheaper, electricity could be auctioned on the Net leading to massive savings, and contracts could be downloaded instantly from the Net.

The first rumblings of what was to bring down the $200 million investment in Wine.com were to be heard from California where old prohibition laws were invoked to ban e-wine sales across state borders.

By November 1999 two million were online in the UK. Today that figure is 10 million and rising. In less than 18 months nearly every household in Britain had connected. This is the story of that year.

Part 1
Pioneers and Turning Points

The first year of e commerce was 99–2000 but pioneers were developing software and elementary web sites from as early as 95–96 and putting them into the unknown void where for many, the expatriate audience was the first sign of trading life.

Top chef introduces revolutionary e-cooking service

One of London's leading chefs, Stefano Cavallini, is pioneering a doorstep delivery of ready prepared ingredients that can be turned into meals within 30 minutes.

Cavallini, head chef at the Halkin Hotel, is widely credited as the father of New Wave Italian cooking in London, a term he calls *Essential Cooking*. He is putting his name to a new cookery service LeapingSalmon.com [http://www.leapingsalmon.com] that pre-prepares all the ingredients for a dish and delivers complete with recipe and instructions for finishing off, each one graded according to the skill level.

Dishes include Roast Breast of Duck in a rich Caramelised Red Onion and Port Sauce with French Beans and Toasted Almonds, Vichyssoise Carrots and Mashed Potatoes infused with Truffle Oil (high skill level); Melanzane & Taleggio Ravioli in a Butter and Pine Kernel Sauce (medium skill level) or Thai Green Curry with Fresh Thai Vegetables and Coconut Thai Fragrant Rice (low skill level).

The dishes are prepared in a central kitchen with same-day delivery promised in London, and next day nationwide.

The idea is the brainchild of two former marketing graduates, James Marshall and Peter Kenyon. Marshall's background is in shipping and logistics and working with international brands such as Coca-Cola and Nestlé, while Kenyon has worked in high-tech industries in the Far East.

Meal kits can be ordered up to five days in advance and cost from £14 to £20 plus £4.50 delivery.

08 May 2000

AC Nielsen launches Web research package

Global research company AC Nielsen has announced the launch of new research into websites that may change the whole industry.

The British version of the new service was launched in May and heralds a new era in the development of the Net. The revolutionary new assessment data will go considerably beyond the conventional page impressions as a measure of the validity of a site. The research is carried out by PC top boxes with java-enabled software which can pick up exactly where the sample audience goes and which banners they may click. Presenting the new package, George Butler said that in the future advertisers were going to be looking for new categories which might even include the surfing preferences of two to six year olds. 'It will be important to know whether they like sites like Disney or ToysrUs,' he said.

The statistics will, for the first time, be available to competitive companies looking to analyse the performance of rival websites as well as their own. It will track precisely who is using the PC, the Net decisions they make in terms of where they go, how long they will stay at a given location: 'It is all about the reasons why people go where they do,' said Butler.

The research will define the age, gender and preferences of a 10,000-strong panel of surfers who have been recruited across the UK. The techniques were developed and are in use in America and are about to be launched out globally. Commenting on an American site FoodTV.com [http://www.foodtv.com], Butler illustrated that although it had a lower hit rate than rival sites, the length of time people spent on each page, and the number of pages each one of those accessed was a better guide to effective Web success. To underline this the site had a predominantly female audience and as such was likely to be an effective advertising vehicle. 'But on this site the evidence tended to be that women were browsing, not buying'.

30 May 2000

Interview: rise of an organic banker

Direct sales from the farm feed the 2000.

Richard Counsell won the Sunday Telegraph/Barclays bank award for e-achievement last week. His tiny site won *Website of the Year*, beating off competition from Iceland and W. H. Smith among others. A former foreign exchange dealer, Counsell swapped a life in the City to go back to his family farm and develop rare breeds, farm organically and use the Internet to develop new business. This is how the Net helped change his life.

'I came back to the family farm in Rodney Stoke, Cheddar, 18 months ago when it became obvious the farm was becoming a big drain and was losing a lot of money. We decided to sell direct to our customers and offer them the very best quality, using the old traditional breeds that have been here on the farm for centuries. The family have lived here for just under 600 years and because of that obviously I was very keen to stay on the farm I love so much. From very humble beginnings we now supply over 2000 people throughout the country with certified organic rare-breed meat.

'Customers order on our online farm shop and receive their order within 24 hours. Because we do all the actual farming, the website has few overheads and I am able to supply better quality meat, and at up to 40% cheaper than any of the major supermarkets/retailers. My competitors have to buy in stock and then mark it up – with Somerset Organics [http://www.somersetorganics.co.uk] you know that everything has been farmed ourselves and as such you are buying direct from the farmer not the middleman!

'Through our net call service, you can speak directly to the farmer (often in the field) through my mobile and be sitting at your desk in central London! This is perfect information and exactly what we hope the customers want. Not only that but each different breed is marked as it is processed, so we are able to tell you exactly what breed you are receiving. It enables our chefs to generate menus such as, *Fillet of Gloucester Old Spot Pork.*

We have recently added a chat forum, there are monthly competitions related to farming, a weather watch, weekly recipes and even a virtual farm tour in the offing, which is a digitally filmed CDROM allowing customers to navigate there own way around the farm.

'This, more than anything else, perfectly illustrates the value of the web to small producers like myself that want to compete on a national level. The goal posts have shifted and the retail market is now no longer controlled to the same degree by infrastructure and cumbersome administration. Somerset Organics and other similar companies have all the advantages now. I am more flexible, have greater access to supply, have lower overheads and costs and because of this in the quality end of the market I, as a farmer, the producer will always be better off than the retail giants. If you can find your market and work hard at keeping your customers happy, then the future for imaginative small producers looks very exciting indeed.

04 July 2000

Pink dot goes PDQuick

LA home shopping pioneer rebrands.

One of the legendary Californian online delivery services has rebranded after 13 years of trading. Pink Dot has become PDQuick.com [http://www.PDQuick.com]. The company offers 30-minute delivery. Chief executive Dan Frederickson said, 'The new name better reflects our commitment to customers to provide just about anything they want in a hurry in minutes. It also represents an important milestone in our 13-year operating history as we position ourselves to dominate this business on a national basis.'

In July, PDQuick received $35 million from its investors to fund its expansion into new consumer markets around the country. An experienced executive team and board members from such well-known retailing and restaurant innovators as Kinko's, Pizza Hut and Circle K are spearheading the company's national expansion plans. PDQuick operates 18 customer fulfillment centers in Southern California and Baltimore, and makes 1.2 million deliveries per year. There is no minimum or maximum order and delivery is $2.99 regardless. The company has specialised in made-to-order everyday meals but also has an increasing number of brand alliances including Campbell Soup, Coca Cola, Ice Cream Partners USA, Jon Donaire, Kraft, McKesson Water Products, McLane Co., Nabisco, Nestle, Orval Kent, Rich's Products, Suiza Foods, Sysco and Vie de France.

14 August 2000

Sales training goes on the Net

Online learning prepares product pitch to 71,000 farmers.

Agway, a farmer-owned cooperative, is using an eLearning solution for distance sales teaching for remote employees. Vice-President Bill Parker said: 'We can reduce the expense and time delays of bringing dispersed professionals together. Based on the dramatic results of this program, we will now be utilizing eLearning in other divisions of Agway.'

Agway is owned by some 71,000 farmer-members in 12 states in north-east America, it produces and markets dairy and livestock feeds, as well as agronomic products and services and had revenues of $1.4 billion in its last fiscal year. The program was provided by the Baltimore-based company Caliber – www.caliber.com. [http://www.caliber.com]

12 September 2000

Pringles pioneers voice mail ads

Potato crisp company Pringles has been trialing voice-mail advertising to support its concert sponsorship and has reported click-through rates of 11.3%.

The research has been conducted in 26 markets across the US. eVoice i-audio advertisements are a new form of 1:1 marketing, in which subscribers hear one 15-second advert after logging into their voice-mail accounts. Those interested in learning more about a concert in their respective area were prompted to click through to receive an opt-in email from the advertiser.

13 September 2000

That Sainsbury's depot . . . how it works

£10 million and 185,000 sq feet and not a customer in site, Park Royal gears up for the new era.

The bakery is just being finished at the new Sainsbury depot to service its e-commerce business. It claims to be the largest depot in the UK and can handle

3000 orders a day served by a fleet of 28 vans. Selection is by conveyor belts and electronically tagged boxes that can grab a tin of peaches.

Sainsbury's e-commerce director Angela Megson is predicting exponential growth with satellite warehouses being added to the mammoth centre. 'Just look at those seven year olds sending text messages on mobile phones, or the growth of e-mail. It is as easy to underestimate how quickly technology will catch on as it is to over-estimate it'. She points out that the average in-store shopping basket is worth £22 where every Internet order averages £100.

23 October 2000

Part 2
Stop Press!

This is the story of the first year of e commerce revealed as snapshots – as they happened – of the hopes, innovations, successes and failures of players big and small.

Beer sales show 300% growth

A specialist beer company is reporting a staggering 300% year on year growth in sales online.

The specialist shop chain BeerRitz [http://www.beerritz.co.uk] first opened its website for business in May 98 claiming to stock the largest range of specialist beers in the world. Sales are exclusively in the UK because despite a massive global demand, export sales have been restricted by the cost of carriage, by carriers not offering insurance on breakages and in the USA by customs restrictions on alcohol sales. 'This is a great shame as the USA represents about 50% of our total demand, which is currently unfulfilled', explains sales executive Yvonne.

Belgian beer tops the list of sellers in general with the *World Champions Pack* being the most ordered single item. 'Many of our customers are very knowledgeable about beer, and they know exactly what they want. We have now built up quite a list of customers who order regularly, some every few weeks. Our profile remains quite low, as we do not spend much on advertising. Most of our new business comes from recommendations and word of mouth, plus general web browsing. Importantly, we are profitable.

'Main glitches are typically associated with people using out-of-date browsers, who run into technical problems while ordering. Our secure server for credit cards works OK and has proved popular. Another problem here is where people placing an order forget to include their expiry date details, etc'.

04 January 2000

Pizza wars online

Domino Pizzas has launched the first salvo aimed at the supermarket share of the lucrative pizza market by going online with direct ordering nationwide.

The Domino pizza site [http://dominos.co.uk] is now offering to deliver pizzas direct to homes within 30 minutes of an online order. The site includes a nationwide map of all its outlets – nearly 200 in all – plus an ingredients guide. Plans are afoot to adjust outlets to take on extra space to service the home delivery orders.

04 January 2000

Hotel bookings service gets off to flying start

A mould-breaking new hotel reservation system has put more than 90 customers into one hotel since launching two months ago.

Laterooms.com [http://www.laterooms.com] is a UK national register of hotels with rooms to let at discount rates at the last minute. Launched before Christmas it already boasts thorough listings in most UK cities, all at discount rates.

First levels of response have been spectacularly high, with one hotel reporting 90 bookings through the service. The hotels can list themselves free of charge for a three month trial period and after that pay a flat rate subscription, provided they offer a discount to Web users who call in using a special code to get the rate.

Director Chris Allen said: 'The problem with the hotels is that they all have different booking systems so it is not practical to book direct via the Net. We had planned to do that but have shelved it. We are staying with the phone reservations system because everyone knows how to use that while many front desks still do not even know the difference between an email address and a website'.

The site is still fairly basic – though a revision is on the way – and posts up a list of all hotels with offers in a given area, with the price and level of discount for each room, often more than 20%. Visitors then click on the offer they want and can see the full details and even, where hotels have a website, even go through to see pictures of the rooms. The booking reference number gives the name of the person on the desk who has the Lateroom.com rate.

16 January 2000

New site takes on Chateauonline

A second major new wine portal launched at Hotel Olympia trade fair.

Wineandco.com [http://www.wineandco.com] is planning to roll out simultaneous launches in France, Germany, the UK and Asia. Initially with a consumer-led site followed shortly by trade prices.

The company is based in Aix-en-Provence and will go head to head against the impressive Chateauonline [http://www.chateauonline.com/].

UK director Seth Lewis said: 'We are taking a different tactic from Chateauonline by working with trading partners to build up visitor numbers. We are also handpicking very special wines right through the price range. The message is 'friendly wines . . . great service . . . passionate people'.

07/02/99

Ex-Burger King team look to supply links

A management team out of Burger King has set up a prototype food service trading site.

Officially launched at Hotel Olympia, the imaginatively named Eggsbenefit.com [http://www.eggsbenefit.com] is looking to provide the link between buyers and sellers on the Net. Restaurants are asked to pay a subscription fee, as are suppliers, and then they are hooked together to develop trade.

The prototype site was only conceived in October and trailed at the exhibition with an impressive stand and boxes of T-shirts. It is the brainchild of Howard Prince Wright who is claiming to have already secured a number of major corporate clients.

07 February 2000

Row over cut price Net rooms

Angry hoteliers have warned rivals of the dangers of getting into a cut price war over room prices.

Speaking at the Master Innholders General Management Conference, Ian Wagge of reservation system Rezolutions warned hoteliers that cutting prices could upset their business. 'Do you want to teach your customers to become discount travellers and teach your rack-rate customers that you are ripping them off?' he told delegates.

The fears of a price war have been sparked with the launch of Laterooms.com [http://www.laterooms.com/] which offers big discounts on last minute booking, also at Lastminute.com [http://www.lastminute.com/] and in America from Priceline.com [http://www.priceline.com/]. Some hotels had asked to use the service but to remain anonymous to avoid the problem.

07 February 2000

Flying noodles get 20 visitors a month

A specialist pasta shop on the net is taking more than half the orders off the traditional business.

The Flyingnoodle.com [http://www.flyingnoodle.com] imports specialist pastas and sauces and sells direct, largely to women in the 35–55 category.

Founder Raymond Lemire says the most successful means of marketing has been via email newsletters. He said: 'We concentrate our efforts on finding the really hard to find products. We taste everything that comes in, before we sell it, because we're pretty fussy about what we sell'.

The main thrust is a monthly order of two gourmet pastas which can be cooked and ready on the table within 15 minutes. Sales tend to follow population patterns, mostly in America, and an embryonic market in Japan fell away when the yen exchange rate to the dollar dropped.

07 February 2000

Herb farm is growing

A specialist herb grower is being approached by four companies a week to supply herbs over the Net.

In all, Jekkasherbfarm.com [http://www.jekkasherbfarm.com] is getting upwards of 2000 inquiries, and that is without having registered the site or developed any e-strategy.

The big breakthrough has been in the last nine months, although the site has been operational for two years. Jekka McVicar warned: 'We are at the moment having about four companies a week contact us asking us to supply them herbs for their website; it is amazing how many of them do not even have a site up and running, it is getting way out of hand. I have agreed to supply 3 sites so far. Each with a different collection of herbs. I have changed from Organic Food Federation to Soil Association as my symbol body, so that, combined with (at last) the interest in organics, means I get no days off.'

16 February 2000

Wine portal looks to open trade account

The massive new international wine portal announced the launch of its trade site with a full roll out from March.

Wineandco.com [http://www.wineandco.com] started life building sites for wine producers from the chateau Beaucastel in Chateau Neuf du Pape, but has now developed into a full-blown portal with consumer and trade dimensions.

It is the brainchild of 29 year old Marc Perrin whose family own Beaucastel. Philippe Faure-Brac, a former world *sommelier of the year* and previously buyer for Air France, has been brought on to the team to select wines from more than 1200 vineyards.

Delivery is guaranteed within seven days, though the London office is hoping to get this down to one to two days. Every one is listed with tasting notes, ageing potential and advice on matching to food. 'Wine should be fun and that is what we want it to be,' said director Seth Lewis.

Backing this up are some web-savvy tricks on the site such as complete vintage and vineyard search engine, wine industry news, a DIY form to ask the wizard-sommelier to recommend wines to suit your palate, a wine encyclopaedia and online forum.

Other countries coming on line soon are Belgium, Netherlands and Japan, with 15 scheduled for phase one of the global launch. The strategy is to grab 20% market share by a combination of marketing and direct link as trade partners with search engines like Virgin and AOL.

15 February 2000

Titan portal nightmare as names clash

Two rival would be portal sites have come head to head with virtually the same.co.uk names.

Marketing nightmare or what? The ambitious consumer site has launched itself as Foodndrink.co.uk [http://www.foodndrink.co.uk]. Meanwhile long standing publisher William Reed, owners of the trade journal The Grocer, has bagged Foodanddrink.co.uk [http://www.foodanddrink.co.uk].

Innocent surfers looking for a nice recipe for watercress soup may find themselves confronted with jobs in the engineering sector, while key executives looking to develop their e-commerce portfolios with paid for advertising space and putting up their catalogue in front of supermarket buyers will be misdirected off into champagne offers and some softie restaurant critiques. Perhaps they will not mind so much. The word *and* itself is coming to be a critical link in website names. Another site to use it is the giant French-based wine site Wineandco.co.uk.

Expect more confusion to follow soon.

16 February 2000

Gourmets showered with gifts

Prototype foodie portal woos surfers with offers and prizes.

Foodndrink.co.uk [http://www.foodndrink.co.uk] has unveiled a series of promotions to lure food lovers to its site. Restaurants are being signed up to accept voucher cards with 10% discounts. The first names are quite small beer, and while the prospect of a free glass of house wine with the main course at Navajo Joe's at Covent Garden may not be in the *Who Wants To Be a Millionaire* category, other names are promised. Other new deals include 30% off food books from Waterstones, a magnum of champagne for quiz winners and a competition to win a meal at home cooked by a top chef. The restaurant listings seem to have hit the usual financial snag of funding and there is a come-on for readers to enter their reviews in return for the chance of a prize.

17 February 2000

Directors savage credit card bureaucracy

The Institute of Directors has issued a damning survey blaming banks and credit card companies for holding back e-commerce.

The Institute of Directors says firms are being deterred from selling goods over the Internet by the reluctance of credit card suppliers to allow online payments. The IoD carried out a survey of its members: 76% of firms responding had customer websites, but only one third of those allowed shoppers to place orders via their sites.

The IoD says this is blocking the development of e-commerce in the UK while the United States is forging ahead. One Internet company, Lastminute.com, says it faces what it describes as 'ludicrous bureaucracy' from the banks.

The organisation representing credit card suppliers, however, says they have to be cautious because they could end up having to compensate customers whose goods don't arrive.

The IoD also called for lower telephone-call charges as a priority over lower line rentals, because members foresaw increasing use of the Internet: 'We must work to ensure that everyone is fully aware, and to remove the obstacles preventing British business from taking full advantage of the opportunities,' said Richard Baron, of the Institute's policy unit.

17 February 2000

Farmers launch spuds online

An enterprising group of Cheshire farmers have launched a home delivery service bringing bags of potatoes to the door.

Roston Partners of Common Farm, Whitegate, Cheshire are now delivering packs of farm fresh potatoes to the door at competitive prices. Orders are placed on-line at http://members.farmline.com/acorns/spuds [http://members.farmline.com/acorns/spuds]. And to get round any fears of credit card fraud (as if) it is good old-fashioned *cash on delivery*.

22 February 2000

Europ@web takes slice of wine portal

Fledgling wine portal wineandco has picked up new funding for its global launch.

The Internet arm of Bernard Arnault's family holding company has acquired a strategic minority stake in Wineandco.com [http://www.wineandco.com]. The investment is by way of a capital increase of approximately 100 million French francs and will enable WineandCo, which already has operational sites in France, Italy and England, to accelerate its international development.

Marc Perrin, president and founder of WineandCo, commented, 'We are extremely enthusiastic about the idea of working with Europ@Web. The numerous operational synergies that can be quickly established will allow us to accelerate our global development as well as giving us access to numerous partnerships.

'Chahram Becharat, managing director of Europ@Web, commented, 'We have been convinced by the quality of WineandCo's management team – who possess a strong knowledge of the wine industry, which is crucial – combined with more than five years' experience of operating on the Internet.'

22 February 2000

Gourmet site picks up IBM award

A bespoke gourmet site has walked off with the top prize for web development innovation under an IBM award scheme.

The Gourmet-connect.com [http://www.gourmet-connect.com] was picked out as a winner among IBM case studies for its low cost total solution to e-business. Managing direct Peter Kirkham said, 'We produce recipe meals and speciality joints and we set ourselves up specifically to become a specialist home delivery/ e-commerce company.

'All our products are freshly produced to order and delivered by overnight dispatch. Our recipe meals have a unique marinade which is coldly miscible with alcohol, our French provincial recipe meals such as Poulet Vallee D'Auge (chicken marinaded in Cider with Sage and Onion) are marinating during the delivery cycle. We also produce speciality meat joints, such as a de-boned chicken leg filled with mince chicken, apricots and a sweet and spicy sauce, tied into a parcel shape.

'We also source from other specialist companies, usually small family owned firms who market their goods locally. Typical of this would be David and Linda who farm 80 acres in Kent, making no money from supplying supermarkets with apples they invested in a rack and cloth press and use hand-picked, tree-ripened apples. They produce an unfiltered juice: when you open the bottle you can smell the orchard!

'The email side has created a lot of interest, we have over 500 registered email customers in this country, and some interest from America, India and the Continent.

'Our marketing plan is to develop a system of agents across the country who demonstrate our products and establish a customer base, at the same time they publicise the website and generally inform/educate about e-commerce'.

22 February 2000

Rare breeds shop goes virtual

Backed by an EC grant to build an impressive butchery in the barn, Heal Farm has set itself up as a virtual shop.

A delighted Ann Petch said the best thing about the web was the high conversion rate of email inquiries to e-customers.

'The virtual shop [http://www.healfarm.co.uk] will contain a whole lot more than a printed brochure, recipes, news, Q&A and links. It will be capable of being updated frequently from here.

'We are quite excited by the response we have had so far, many more e mails than phone calls from the mention in the Sunday Times supplement, and the e-mailers converted quickly in to customers. Immediate sort of people.

'Unlike many of the new e-commerce companies we have both the fulfilment and admin capability to respond very quickly to new orders. Our new building will easily cope with a lot more throughput. Supply is no problem as several of my rare-breed chums would love to produce for us, to our high spec of course.'

22 February 2000

Those 25,000 Unilever jobs – the real story

The job losses made the headlines but the Anglo-Dutch mega corporation was putting 20% on its procurement online and investing 200 million Euros on the Net . . .

The cost of the change however was what made the headlines: 25000 jobs to go worldwide. Most of the jobs to vanish will be in Europe and the US, the two areas where penetration of the Internet has been greatest.

As part of its restructuring plan Unilever will spend 200 million Euros this year alone on e-business initiatives: it will put as much as 20% of its procurement online over the next three years; it will step up its buying through online auctions, which it has already piloted in America; and it will work on an existing joint venture to sell directly to consumers over the Web.

26 February 2000

Remote poultry breeder gets 70 visits a day

Kintaline may seem like some high-tech jargon but it is a smallholding eight miles from Oban, in Argyll.

To supplement her free range egg business, Jill Bovis has set up Kintaline Farm Plant and Poultry Centre [http://www.poultryscotland.co.uk/], a resource for information on rare breeds of chicken and duck and acting as a conduit for fertilised egg sales around the world. The site is getting 70 to 80 visits a day and has inquiries from Australia and America.

26 February 2000

Spicing up the net

Sunday Times mention brings 820 enquiries.

Specialist spice shop Bristol Sweet Mart [http://www.sweetmart.co.uk] received 37 orders the morning after a mention in the newspaper, said owner Abdul Ismail. His father set up the Bristol-based shop 20 years ago as a refugee from Idi Amin's Uganda with £50.

'We sell 52 varieties of chili sauce alone, from Mexico to Malaysia. So far we have about 10% of our total stock online. The kinds of people who use the site tend to be professionals, most of them from London. The site is for people who love ethnic food, not so much the ethnic communities themselves.

'We have had the site since spring 1999. It was an easy-to-use, simple, basic site. We had very little luck to start with. Business has picked up and we are in the process of revamping the whole site. The net business will remain second to our wholesale and retail business since the orders can be very thick sometimes and very thin at others.'

26 February 2000

UK launch for wine auction site

The successful wine auction site which now employs more than 30 people in America opens a UK site, having poached top Christie's wine specialist.

Winebid.com [http://www.winebid.com] was set up in 1995 by Chuck Parsons, an American wine collector looking to sell some of his 10,000 strong collection. Now its twice-monthly auctions in America have a regular 8000 bidders.

The first UK auction starts on March 16 and will run for 11 days. Bidders are notified by email if they have been outbid or can autobid up to a specified amount. All the wines are vetted by Winebid first and Chris Munro has joined from Christie's where he was the European Sales and Senior Wine Specialist. Global wine auction sales are dominated by Christie's and Sotheby's and last year had a turnover of £64 million.

26 February 2000

Case study: espresso is hot (part one)

Redundancy inspired Curtis Claar into a new web career and a profitable business.

Espresso.com [http://www.espresso.com] was born as an email address in a would-be newsletter for coffee lovers out of Seattle even before the Web was really running. But the moment the site went up in February 1998, it was an instant hit.

Claar recalls: 'WOW! was I impressed! With NO MARKETING at all, the traffic was well beyond 1500 visitors a month instantly. Now that we've marketed it and have made it into something fairly worthwhile, we are amazed each month with the increased traffic. In fact, we're now able to sell advertising (to people like Nestlé, Seattle's Finest, etc.) and it's actually a worthwhile venture.

'Before we launched The Espresso Index, people were offering thousands of dollars for the domain name – we're glad we didn't sell out. There's no telling what it would have become.

'There's still a lot of work to do on the Index, but the original goal of providing something useful is still our goal.

'Since the domain name was so popular, we even launched EspressoMail, which has been fun too. This gives people a chance to at least use it as a part of their email address for a low fee. More on that topic at: http://mail.espresso.com [http://mail.espresso.com].

'We've also launched the Espresso Forum [http://www.espresso.com/forum], and rolled off the Espressop50. We recently acquired espressorecipes.com [http://www.espressorecipes.com] as well.'

09 March 2000

Case study: espresso is hot (part two)

Consumer tests trigger the big hits for the hobby site that became a world leader.

Web author Mark Prince has grown his hobby site into one of the cornerstone sites for coffee in America. In under six months it has grown from a new registration to getting 2,500 visits a day with an average surfing time of more than 10 minutes.

This is the Prince philosophy: 'Well, I say it right on the front of the site: I'm a slave to finding the highest quality espresso and coffee I can make or buy. It's a passion of mine, though my daily intake is pretty low. But about a year ago, I was getting bored with my site and while surfing the net trying to find quality coffee or espresso sites, I was a bit surprised and sad to find that none really existed. Some contained a LOT of content, but really poor design and it was hard to find info. Others were OK designs, but light on the content. And a year ago, I could find absolutely no information based on my two favourites – Rancilio espresso machines (only found commercial listings), and vacuum brewers (found nothing!). So I decided to do it myself.

'Coffeekid.com [http://www.coffeekid.com] was originally part of my personal site, but it quickly outgrew it, so I registered the domain name (October 99) and redesigned it from the ground up. At first, it contained all the main sections except the review section. It was mostly my own ramblings, thoughts, theories, pictures, etc. etc., and guest articles in the Rancilio section.

'I tried to add stuff to the site every few days (still do), and I tried to maintain a level of quality in the site, from the pictures I used (almost every photo on the site is mine, ditto for other images), to consistent layouts, menu systems, and navigation. At one point I wanted to add a site map, but the site is now so expansive (some 200+ pages and about 400+ graphics) it became a challenge I didn't want to do. I had the rather unmodest objective of wanting to build the best darned non-commercial coffee website on the Net, and (again, unmodestly), I think I've done that.

'The site generates about 25 'fan emails' a day to my inbox (not including review site email), and it makes me feel pretty good that people are now able to find the kind of site I myself was looking for a year ago.

'On the subject of the review component of the website, in December 99, in the newsgroup alt.coffee [news:alt.coffee], people were lamenting the lack of a good online source for reviews of coffee and espresso equipment by consumers, or at least by folks a bit more unbiased than the stores that sold the same products. Someone brought up deja.com [http://www.deja.com] and how they have some ratings and reviews of some espresso machines, but those reviews and ratings were not taken seriously. I talked to my chief programmer in December about us possibly setting up such a thing. We decided to code it over the Christmas holiday.

'I wanted to do it primarily to provide a public service, one that was needed (as I said, no real online review service for this speciality equipment existed prior to mine), and also to provide a place for my fellow alt.coffee contributors to post their thoughts on their machines once, then point people to the review next time someone came into the newsgroup asking 'what do people think of product xyz?' We also did all the custom work with the intention of developing the software, the database components, and other assorted programming for future selling through my company (repackaged, customised, of course).

'We invested over 200 person-hours in the development of the code, without any income or payment for it, but we aren't stupid: If we can impress our clients with the programming of the review database and components, we can probably resell it (not the info in our reviews, but the programming code).

'So we launched the review component in early January, and only with advertising in alt.coffee, it was a hit. I was getting 500–1000 unique visitors a day to the site (before the review component, coffeekid.com was getting about 300 visitors a day) within the first week. Then someone at Yahoo heard about the whole site, and listed it as a Site of the Week, and it took off. I went from

1000 visitors a day to 7500 or more a day, with an average *stay* time on the site of over 10 minutes, and in February, we transferred some 12Gb of data from the domain down the Internet pipe. Even today, our sustained daily visitor count is around 2,800 a day.

'The site has been written up in magazines in the US, Switzerland and even New Zealand. It's getting exposure on sites like Salon, MSNBC and other popular sites. Because of the popularity of my little hobby site, I have to face something I didn't want to face – my site host wants to bump me up to a higher bandwidth pricing category, thus changing my cost of running the site from $30Cdn a month to about $150 a month.

'Because of this, I have to start thinking about running a few ads. So my plan is this – I registered the domain name www.coffeegeek.com and in May, I will move the review component completely over to the new domain (and redesign it), and keep the rest of the coffeekid.com content and design right where it is now. I will start advertising on the geek domain, but keep coffeekid.com free of all ads. I could run banner ads (I've been approached about 2 dozen times by people wanting to *partner* and advertise), but I don't want to for ethical and philosophical reasons – coffeekid.com is my personal hobby site, and I simply do not feel right running ads on it. And I never will. CoffeeGeek.com? Yeah, since that will be a purely *public service* website, I will run minimal ads (I'm going to try to run only coffee product related ads), but that is down the road.'

09 March 2000

Fishmonger delivers within 24 hours

A traditional fishmonger with his own inshore boat can deliver almost on the same day.

Fifty nine year old Jeff Martin has handed his fishing boat over to his son Chris to concentrate on the family shop and new Internet business from Newquay in Cornwall.

'I don't see how any of the supermarkets can be selling fish less than seven days old,' he said, 'Technically, if fish is kept at the right temperature, it can be in prime condition up to 12 days after it is caught. But if the temperature goes wrong, it deteriorates rapidly.

'We are buying most of ours straight from the boat, putting it into a container where it remains at optimum temperature and delivering it to people the next day.'

Delivery charge is £5 and Internet orders attract a 10% discount. The site [http://www.martins-seafresh.co.uk] was designed by Martin himself using Front Page 2000. 'We are a small family business and for us, this is really just extra trade at the moment. But in the long run I can see it being bigger than the shop.'

12 March 2000

Virgin snaps up orgasmic wines

After records, planes, trains, lottery Richard Branson has moved onto the Net with wines.

The new site promises the usual clichés about demystifying wine and how supermarket wines are bland and dull, hardly an opinion held by most masters of wine. Sir Richard, whose Virgin Group owns 50% of Virginwines.com [http://www.virginwines.com] said: 'I find it ironic that the UK has the largest and most cosmopolitan wine market in the world with some of the world's most innovative wine merchants and yet the majority of the population settle for bland wines sold in supermarkets.'

It aims to become the UK's largest online wine stockist, selling up to 8,000 labels at discount prices under Rowan Gormley, the South African who turned Virgin Direct into a £2.5bn business with 300,000 customers. Wines will be classified by taste rather than name or region, reducing the knowledge required by customers. Experts will write online wine reviews and customers will be encouraged to post their own comments. As an entirely online operation, Virginwines believes its low overheads will allow it to compete against supermarkets and online services being developed by *bricks and mortar* wine merchants.

Virginwines emerged from Orgasmic Wines, founded in October by a group including Laura Knight and Clayton Gormley, Rowan's younger brother. The owners of Orgasmic will own half the new company. Ms Knight said: 'Only a few sites will survive but with the strength of the Virgin brand we are determined to be one of them.'

12 March 2000

Food from Britain launches corner shop for small producers

The mall-style site allows member companies to enlist for as little as £199 and claims even a Net novice can get it up and running in two hours.

Speciality-foods.com [http://www.speciality-foods.com] is also promising a links strategy into bigger sites like Handbag.com [http://www.handbag.com/], Richclickings.com [http://www.richclickings.com/] and Charlottestreet.com [http://www.charlottestreet.com/] to drive traffic. It claims to be a 'virtual high street'. Among the first companies to open for business are Duddleswell organic cheeses, Alderley Trout from Gloucestershire, and there's even English wine from Cornwall. Director Charlotte Lawson is further negotiating a 12.5% discount on handling global deliveries with ParcelForce.

15 March 2000

Another kite mark to validate e-business

The scheme has the backing of the DTI.

E-commerce minister Patricia Hewitt said: 'Fears of insecurity and fraud are holding online consumers back', a statement which rather flies in the face of the exponential growth in e-shopping. The TrustUK mark requires companies to ensure their advertising and contracts are fair and balanced, their shopping systems secure and a privacy code is honoured.

Applications are available from www.trustuk.org.uk [http://www.trustuk.org.uk], and in return companies get an e-hallmark to display on their site. The accreditation programme will be overseen by Lord Borrie, former director general of the Office of Fair Trading. It launches in May.

15 March 2000

Domino sells 6000 pizzas a month

The pilot pizza home delivery from Domino has taken over 2% of total turnover . . . but not without a few glitches.

The Milton Keynes-based chain [http://www.dominos.co.uk] has stolen a march on competitors with home delivery direct from the Net or through Sky TV. The project is selling 6000 a month nationwide. Staff were slow to pick up orders sent by email and fax, and now the system has been replaced by a new digitised central system with screens positioned over the pizza makers' heads.

The Internet delivery service has seen a rise in people wanting to create their own. 'We are not talking about huge numbers yet. But we did get a 10% increase last month on Internet ordering. We are definitely moving in the right direction,' says marketing manager, Anna Taylor-West.

16 March 2000

¡Olé! Here comes Spain!

A small company has set up a specialist site dedicated to Spanish foods.

Director Chris Gayton has been test-marketing his product range at Spanishfooddirect.com [http://www.spanishfooddirect.com]. He said: 'We are only just starting to attract any reasonable number of visitors, for example 68 on Sunday and 48 yesterday, etc. We are now starting to collect some feedback from visitors.

'We started with a basic range of typical Spanish food products however there has been little actual trading so far, one of the problems being that to make these products economically viable they need to be sold in units that are probably too big for the average buyer: for example a whole Serrano ham.

'To some extent this was expected and we are now using the feedback that we have to examine the viability of other products that may suit the Internet/ mailorder style better, although the combined factors of buying in small quantities and the cost of onward delivery to the customer make this fairly tricky to do and make a profit as well. However, we regard this as a long term project and will keep working to make the site viable in the future'.

28 March 2000

Chocolate orders: five to one now off the Net

Topline chocolate company Valhrona is now getting five times the orders via the Net compared to traditional mail order.

The long-standing site, set up in 1997 with the much coveted url chocolate.co.uk [http://www.chocolate.co.uk], has seen a surge of e-commerce in the last six months with a peak at Christmas.

E-commerce director Nicola Porter said: 'Junk mail used to be a big problem, especially with orders from Korea and the Far East. Also levels of anoraky-style technical questions or just chasing down customers who had not filled in the form properly, especially in far-flung places. But since last September there has been a notable increase, and e-sales now account for 10% of total turnover.

'The biggest orders tend to be for five to six kilos at a time which are between £180 and £250. Mail order traditionally is a spend of around £55 while the Net gift area is much smaller, but dynamic, at about £20. We get about five to seven queries a day now, often quite technical questions about the nature of chocolate. To date we have had one incidence of credit card fraud from Rumania which cost us £800'.

29 March 2000

Farmers undercut supermarket prices

A dedicated online deliverer of fresh vegetables is claiming to be able to offer foods not just fresher than a supermarket, but cheaper too.

Veg-net [http://www.veg-net.co.uk] is owned by Farringtons, a farming company at Anchorage Farm, Hesketh Bank, Preston, Lancashire. The core activity of the farm has been the growing and selling of vegetables to wholesalers and supermarkets, with Farringtons owning an operation that produces 60 million plants for local farmers.

Veg-net launched in August 1999. Director Paul Farrington said: 'We have just undertaken a new strategy to get more custom, apart from having ultra fresh fruit and veg. We now claim to have the cheapest prices, and ordering from us would be cheaper than going to the supermarket. Plus you get home delivery.

'The ups of Net trading are that as a farmer we are getting a better price for our products but also we are supplying the consumer with ultra fresh vegetables and fruit in top quality condition, one thing to a degree we cannot control when sending to the supermarket. The down is that it is a relatively new concept and people are all shopping in supermarkets at the moment; hopefully this will change.'

07 April 2000

Computer company wins the free phone race

Forget AltaVista, forget ntl: the first totally free Internet access has been delivered by a computer company.

From March 17 all Evesham microcomputers [http://www.evesham.com/index.asp] were supplied with free, un-metered 0800 Internet connection as standard . . . quite a fillip to new computer buyers even while the headlines have been hogged by portals.

25 April 2000

Diet foods boom on the net

Three and a half million bars of diet chocolate expected to be sold this year.

Brian Smith, chief executive of the Pasadena-based company said: 'Synergydiet.com [http://www.synergydiet.com] is exclusively importing gourmet low and zero carbohydrate foods from manufacturers in Australia, New Zealand, Mexico and Brazil, sweetened with Splenda(r).

The U.S. market for low carb foods is sizzling due to the low carb diet craze started by the Atkins Diet, Sugar Busters et al. and the increase in diabetics world-wide: 100 million diabetics in the world (16 million in the U.S.) and 300 million projected by 2002.

'A chocolate bar that is manufactured by Darrel Lea in Australia sold only 80 cases a month in Australia. By October 2000 Synergy will be selling 11,800 cases (over 3.5 million bars) per year in the US via business-to-business (B2B) and retail (B2C) entirely over the Internet.

Americans consume $23 billion of chocolate and confectionary a year. SynergyDiet will be developing specific websites for children, zero carb foods and non-Splenda(r) based low carb foods as well. Their other sites include LowCarbTreats.com [http://www.lowcarbtreats.com] and ZeroCarb.com [http://www.zerocarb.com].

08 May 2000

Phone charge rip-off continues

The eFoodNews.com pick of this week's 'free' services.

Still no sign of AltaVista's mould-breaking free phone service, with the only genuine free Net access still available, as revealed here by eFoodNews.com, through buying a computer from Evesham.com [http://www.evesham.com/index.asp].

The best deal to date remains LineOne [http://www.lineone.net/] with a £20 charge for a box and a flat £5 a month, or for this month ic24.net [http://www.ic24.net/] – free copies from Esso garages – with free calls at weekends and evenings.

Compare that to BT's best deal of £110 a year, although the monopolistic giant is promising new deals for June including a £360 a year flat rate for using the Net as much as you like.

Questioned on its rapacious pricing policies BT just falls back behind a mercenary *we-just-sell-you-the-line-rental* approach. The Internet arm is allegedly a different division. While it prevaricates, other sharks move in like OneTel.Net, offering the Internet for just 10p a minute. Great: that's £6 an hour, or £48 a business day, or a whopping £17,742 if you left it on for eight hours a day all year . . .

08 May 2000

Net trade persuades DHL to float off online services

The massive rise in e-commerce has led to dizzy speculation the company may be worth $1.5 billion.

DHL operates 70 websites worldwide, 40% of its business is online, and 90% of its parcel tracking is now done electronically. The UK site Redplanet.com [http://www.dhl.co.uk/index_js.htm] is claiming 110,000 visits a month, twice last year's figure.

Known mostly as a parcel delivery company in the UK, in fact the company is an airline in the USA and operates across 227 countries, employing 63,000 people and flying 200 planes.

15 May 2000

Whittards looks to brand alliances

Specialist coffee shop Whittard of Chelsea has lined up a series of big brand deals designed to support the re-launch of its site in July.

Chocolates from Bendicks, jam from Wilkin of Tiptree, Walker shortbread and Portmeiron china are just some of the names to be added to the portfolio [http://www.whittard.co.uk].

Chief executive Will Hobhouse said: 'So far, 70% of our Internet sales have come from the US. We are hoping to generate £1 million in sales in our first year. Delivery will be free for orders over £20.'

It is reported the company has spent more than £800,000 on its website and software.

15 May 2000

Coffee Republic signs up with Net cafes

The coffee bar chain has announced a tie in with Simply Internet which own 65 cybercafes.

The first joint venture has just opened in Manchester. Plans for franchise across Europe are under development. Simply Internet chairman Robert Proctor revealed that operators were expected to buy in the computers from his company and then received a rental of £50 a month per monitor in return. Similar deals are in the pipeline with Starbucks, AMT, Madisons and Costa Coffee. While Easy-Everything, whose cafes are claiming between 5,000 and 6,000 visitors a month has a deal with Nescafé.

The Victoria branch is so big, with 358 terminals, that surfers have to order their coffee online.

15 May 2000

Cyber slimmers clean out the larder

A Bolton-based diet programme has come up with a canny Net scheme to help people slim: no other food is allowed in the house.

DietConnect.com [http://www.dietconnect.com] deliver the next day from an online menu of hampers which are calorie-controlled and personalised. The meals even include fruit, vegetables and a bottle of wine for two nights a week. The average spend is £240 a week, but dieters are only allowed to eat from the hampers. Dr Brian Newman who devised the service at his practice in Bolton allows patients to eat up to six times a day and claims some patients have lost as much as 6 kg in a month.

15 May 2000

And not to be left behind ...

Majestic packs up a case of Net solutions.

The increasingly crowded mainstream battle for a share of the online wine market had another player this week ... Majestic wines [http://www.majestic.com]. A relaxed Majestic boss, Tim How, immediately declared his allegiance to bricks and clicks, but then pointed out that the company already has 250,000 regular mail order customers. 'And each of our stores has skilful staff, large stocks and importantly, at least one van'.

Two other potentially critical advantages are the offer of free delivery and the traditional offer of mixed cases. Majestic currently has 75 stores nationwide.

15 May 2000

Jamie Oliver splashes out on his name

TV chef Jamie Oliver has splashed out to buy his own name amid rumours that he is about to move on to the Internet.

Oliver has reputedly had to pay £4,000 for his co.uk name and another £20,000 for jamieoliver.com. A team of funders and advisers is in place looking to explore the options for the young cook who is fast catching Gary Rhodes and Delia Smith as the most popular of celebrity cooks. 'We are examining all options,' said one adviser.

30 May 2000

Marketing

The Net development has helped boost a six-fold increase in sales in the last 18 months for a niche gourmet food company.

Keith Pordum, managing director of BonAppetitDirect.com [http://www.bonappetitdirect.com] has used the website to back up his door-to-door delivery through a team of sales agents. The company set up in 1995 now has sales of £4.12 million and a sales team that rises to 250 in the summer, selling gourmet frozen foods in trays of eight to 16 portions at an average of £25–£50 an order. He cites the main reason for the development into e-commerce as a means of supporting and developing the door-to-door sales force.

'There are three main reasons why people don't buy off the Web,' said Pordum. 'They don't trust you. They are worried about credit card abuse. And they don't know who you are.

'Customers are used to being sold to. On the Web there is no one there to close the sale, which is why the statistics in general show so many sales trolleys are dumped at the last minute.

'Customers need to know what's in it for them. So we incentivise them with offers.' One of these is a free freezer with orders over a certain level. This has successfully created a key, loyal group of buyers.

30 May 2000

Barclays launches e-business payment scheme

The Bank says it can en@ble a business into online transactions in days now.

Other facilities, which have been beyond the reach of small players, include taking payment via Mastercard, Switch, Visa and JCB. According to Barclays' [http://www.barclays.co.uk] figures less than 32% of companies with less than ten employees have web sites and hardly any of those that do are able to accept payments on credit cards.

The bank has also set up Barclays B2B.com [http://www.barclaysb2b.com/], an electronic trading network for financial and business services.

30 May 2000

Banner advertising flops

New research has confirmed that banner advertising on the Net is not an effective means of developing traffic.

Launching its new research programme assessing the Net preferences of a sample of 10,000 surfers, George Butler of AC Nielsen revealed that even banner sites on major portals were failing: 'The click-through rates for banner advertising on these sites is abominable. On Amazon we are looking at some banners with rates of less than 0.1%.'

In comparisons between different industries though the food sector fared particularly badly. There is some evidence to show that on car sites, the click-through rate might get up into the two and three percent level, but on a random sample of American food sites levels were frequently below 0.1%.

02 June 2000

One million new users sign online in April

30% of the UK is now on the Web, according to AC Neilsen research.

And around 8 million are active web users. During April, 62% of web surfers were men. While in America men and women log on in equal numbers.

We spend an average of just over five hours online a month, compared with nearly nine and a half hours in America. Microsoft and Yahoo dominate the top five internet domains. Top of the most popular domains was MSN.com, which had three million visitors. Yahoo had 2.4 million users. These two were followed by Microsoft.com, Yahoo.co.uk and Microsoft's Passport.com which is linked to its Hotmail email. The top British-run website was the BBC, which came in at number nine.

02 June 2000

Not so free after all

What BT fails to tell you about the new Surftime.

The slumbering monopoly of BT's phone service may seem to be about to wake up to the Internet age with the arrival at last of Surftime: a £19 a month flat rate for all Net usage.

The catch is that it is only available to those ISPs that have signed up to BT to take a transactional cut. This may or may not signal BT's move to grab a share of the new ISP market for itself. Executives fall over backwards to point out that phone services are separate from the Internet arm. The public monopoly seems quite happy to carry on exploiting customers by making it as hard as possible for UK surfers to get any deals at all for Net usage.

Meanwhile a survey has shown that 46% of UK users would use the Net more if calls were free as they are in America. This has caused panic among ISPs and many of the better-known brands are maneuvering behind the scenes to work out their response. But with nearly half Freeserve's £7 million turnover coming from its cut on calls, it is not quite as free an Internet service as it claims.

07 June 2000

Le Franc re-brands with Henrietta Green

Gourmet mail order specialist Marchentlefranc.com has re-positioned itself with a new name, new endorsement and new look.

Out goes the cumbersome marchentlefranc.com and in comes Marchents.com [http://www.marchents.com/], a mail order specialist with Internet ambitions. The current portfolio of 300 lines is promised to expand to 1000 by September. Close links with broadcaster Henrietta Green are being touted. Green 'works closely with our head buyer' and Marchents.com is also taking a stand at one of her Food Lovers Fairs.

07 June 2000

Whoops! *I Luv* delays chef's launch

A reputed £1million investment with a galaxy of star names, but it all counts for nothing when the developers think someone loves them.

The launch of Foodoo.com [http://www.foodoo.com/] has been set back by six weeks by the Love bug that hit its development company Oyster.

Big names who have signed up in the hope of a share of the profits include chefs Gordon Ramsay, John Turode and Gary Rhodes. Ken Hom has been snatched from Chopstix.net. They have all supplied ghostwritten recipes to the project in return for a profit share.

The site hopes to make money from advertising and sponsorship, sales of books and cookery items.

07 June 2000

Mail order now taking half its orders on the Net

In the space of two years a mailorder chocolate firm is now getting half its business off the Net.

HomeChocolateFactory.com [http://www.homechocolatefactory.com/] sells kits to make chocolate at home. Originally started in 1996 it first went on the Net two years later. 'We get around 50% of our customers through the Internet, although we don't have online ordering, said a spokeswoman.

'There are no big differences of what people who come through the Internet order, but we do get a lot more enquiries from men. The best thing about the Internet is that you can inform people of special offers, discontinued lines, etc. without the cost of a mail out.

08 June 2000

And what wine would you suggest for turbot, monsieur?

French site sets up online information service with e-sommeliers.

ClickOnWine.com [http://www.clickonwine.com/] will offer direct access into wine chateaux for wine drinkers across the world. The site launched in France with 500 bottles for sale and hopes to bring this up to 1000 in time for the UK and German launches later this year. Beyond the usual books, surfers can also take shares in vineyards and book wine holidays.

'Our aim, contrary to many other sites which exist today, is not necessarily to sell,' said Alain Gannée, master wine waiter at Paris restaurant La Marée and one of the consultants working for the site. A chain of cyber wine bars is only on the business plan.

08 June 2000

Chocolate tops the sales in mall

A specialist chocolate company has used the Net to expand the brand worldwide and picked up one corporate order of more than $7000.

Harry Hall, head of EnglishHall.com [http://www.Englishhall.com/] has been developing Ackermans chocolates [http://www.Englishhall.com/ac/ac/achome.html] through his mall since last September. 'It is the top selling company on EnglishHall.com, the only site to stock the full packaged range', he revealed.

'Individual purchases boomed over Xmas 1999 (we launched in September) and we also generated a corporate order worth over $7000 in November. Marketing has gone well: we have received two mentions in the British press and a favourable review for our packaging and delivery on Shopspy.co.uk [http://www.shopspy.co.uk/index.htm].

'The favourite choice is the 'classic' gold box assortment, available in three different sizes. Overall the company are delighted with the way Internet business has gone: it gives Ackermans the world-wide exposure they were lacking and brings an exclusive, luxury chocolate brand to the world market for the first time. Ackermans only have one store of their own in London so the impact on total sales has been large. The shop still currently sells more but this may change in the near future. We have made minor alterations to the site and are hoping to get new photos done of all products before Xmas this year.'

08 June 2000

Seattle coffee profits head into cyberspace

The sale of Seattle Coffee Company to rival Starbucks netted Scott Svenson and his wife Ally £20 million which is now being used to back Internet food ventures.

The couple have been revealed to be the driving force behind investment arm Eos run by ex-Excite guru Jed Simmons. Eos is funding the new Foodoo.com [http://www.foodoo.com/] site which has signed up star chefs including Gary Rhodes and wine buff Hugh Johnson. When asked Svenson couldn't name the chefs his company had signed but responded: 'A lot of things play off each other'.

He also has shares in a series of cafés with Antonio Carluccio. 'It's a real holistic thing. If you do something because you are passionate about it, success and the financial stuff takes care of itself'. A comment which did not stop *The Times* referring to the Foodoo.com site as 'half cooked'.

20 June 2000

Survey slams online services

Four out of five UK Net companies flunk the e-tail test.

Lack of security and not minding the small print were two of the criteria in a survey by Trust On-line. Companies were also criticised for poor delivery, poor returns policies and not giving proper contact support. Sainsbury's, Tesco, Harrods department store and even Virgin and Interflora all failed to meet the standard set by Trust On-line. The 50 leading UK e-commerce sites were tested. Forrester research in America has recently shown that some Tesco customers wait up to a week for a delivery slot and see 13% of each order filled incorrectly.

New economy names fared no better as lastminute.com [http://www.lastminute.com], Letsbuyit.com [http://www.letsbuyit.com], petspark.com [http://www.petspark.com] and the now defunct boo.com failed to meet the standards.

Lastminute.com had poorly stated terms and conditions, and returns policy while letsbuyit.com lacked full contact details and a statutory rights clause, according to the survey.

03 July 2000

UK market matures

Survey shows how the Internet shopping market is reaching critical mass.

The latest round of UK Internet User Monitor from Forrester Research shows ABC1 (upper, middle and lower middle class) users now constitute 74% of the user population, down from 86% in December 1998. Use of the Internet by blue-collar and manual workers in the C2D (skilled and other working class) social group has almost doubled to 24% of all users since May 1999. About 15% of all C2Ds are now online, in comparison with 39% of ABC1s.

Half the country aged between 15 and 24 now go online regularly. This time last year, only 30% of people in this age group used the Internet. Older teens and young adults now constitute a third of the UK Internet population, although they only make up 12% of the general population.

Just over 80% of Internet users go online at home while 41% log on in work.

03 July 2000

UK grocery market now bigger than the rest of Europe

Watch out as the big players start to diversify.

American research company Forrester is predicting that the UK model will overtake the American experience before 2005. 'Today, Europe's food retailers only experiment on the Internet – serving small groups of customers out of existing stores and booking 0.1% of sales. But they expect online sales to grow 20-fold in the next two years alone,' explained Matthew Nordan, senior analyst for Forrester Research. 'Online sales will become the food retail industry's key battleground. Those that dominate will diversify to overcome industry turmoil, while those that don't will struggle.'

Aggressive plays from Tesco and Iceland drove a 99 million online market last year, bigger than the rest of western Europe combined. Forrester forecasts that 7% of the UK's grocery sales will go online in 2005. The Nordic countries will follow closely at 6%, while Germany and France will overcome early barriers to reach more than 3% of sales online in 2005. Cultural factors and low technology penetration will constrain growth in southern Europe.

Forrester is predicting that well-positioned food retailers will use the Net to reinvent their businesses. They will use online sales to both strengthen their existing competitive position and decrease their dependence on cut-throat grocery products by diversifying into non-food products and service online. In the report, Forrester developed an expansion matrix, which identified four categories of products and services that contribute to revenue and profit.

'Only a few food retailers will be able to execute this diversification strategy on their own: those with an early-to-market online presence, a powerful, elastic brand, and a demonstrated commitment to online sales,' explained Nordan. 'But late-to-market entrants must retaliate by offering better value, hiring logistics knowledge, and partnering with stronger brands for breadth.'

For the report *Online Grocers Diversify*, Forrester interviewed executives at 40 food retailers evenly balanced across Western Europe. Five were online pure plays; the remaining 35 were traditional bricks-and-mortar food retailers with an average of 868 stores each. Of these, 54% already sell online today or are running pilots, and 77% will sell online by year-end.

03 July 2000

Tesco tests Sunday trading laws

24/7 may be the catchword for e-commerce but it also contravenes historic bylaws.

Currently stores of more than 3,000 sq ft are only allowed to trade for up to six hours on Sundays, but Tesco has extended its home delivery service from 10 am to 10 pm in 35 stores. As a concession, the stores will be closed to customers, but open to pickers for online orders.

Meanwhile Tesco has also been saying that it wants to have 45% of its trade overseas within two years, mainly in eastern Europe and Asia, which may allow the new Internet arm to leverage new markets and tactics.

04 July 2000

As if the web did not have enough recipes . . .

More chefs leap on the Net bandwagon.

Following the launch of Foodoo.com [http://www.foodoo.com/] with top line chefs like Gary Rhodes and Gordon Ramsay, along comes Recipeweb.co.uk [http://www.recipeweb.co.uk/] headed up by Savoy chef Anton Edelman and Belgo supremo Richard Coates.

The tenor of this one is scampi with coriander pesto, and promises a database fully searchable by calories, ingredient, cooking time, cuisine. There is a *culinary counsellor* and the promise too of video recipes in the future, plus competitions with offers to attend a cookery course at, surprise, the Savoy.

The commercial emphasis appears to be towards advertising. Any takers for the squid page?

04 July 2000

Priceline to take on Europe

The cut-price Net-shop and -grocer that allows customers to name their own prices is moving into Europe.

Priceline [http://www.priceline.com/] claims to be the best-known internet company in America after Amazon.com. Its large-scale advertising campaign has featured the actor William Shatner, Captain Kirk in Star Trek, and this year the business expects to turnover £660 million.

Priceline users name the price they are prepared to pay for goods ranging from flight tickets and new cars to groceries and fuel. Sellers then have to match those prices if they can. The new firm, Priceline.com Europe, will be headed by Denis Malamatinas, formerly chief executive of the fast food chain Burger King.

'The sheer intrigue of building a leading e-commerce brand in Europe was too tempting to turn down,' Mr Malamatinas said. 'I've spent my entire career building brands and Priceline is a very powerful proposition.

'For the first time buyers and sellers are able to connect directly. Consumers can regularly save 30-40%. We believe Europeans will respond favourably and quickly to this pricing model.' The initial European launch will be in the UK, moving into Germany, France and Benelux in the following months.

The European site will offer customers the opportunity to name their price for telephone calls, airline tickets, hotel rooms and car rentals. A reply is promised within one hour.

'There are 450,000 empty seats on planes flying around Europe every day,' he said. 'It's like 900 planes floating around the skies with no-one onboard. Airlines are desperate to fill those seats.'

Since the service began in April 1998 more than 5.3 million people have used Priceline, generating revenues of £207 million in the first quarter of this year.

04 July 2000

Over 55s drink 15 bottles of wine a week

Survey reveals Internet drinking habits.

A national survey by WineAndCo.com [http://www.wineandco.com] has found some surprising statistics, including one that says that one in ten over-55s drinks more than 15 bottles a week.

Southerners prove to be the stingiest when it comes to purchasing their favourite vino, one in five of those in the Meridian region spend less than £3 per bottle while West Country wine drinkers are the most likely to splash out, but still only 7% are willing to spend over £10 per bottle.

Scotland and the Central region are black holes where one in three people do not drink wine at all.

04 July 2000

. . . Which is why Richard Branson moves in

Virgin has spread its burgeoning Internet interests wider still with a move into the wine sector.

There is a cute drawing of Richard holding a glass of wine, strange some may say from a man who never used to drink. Virgin is offering 17,500 different types of wine and provides the now very familiar slogan that it will de-mystify wine for the common man and woman.

Laura Knight, managing director, said: 'How many times do you go into an off-licence and look at the shelves, not knowing what to choose? Our site has a *wine wizard*, which gives one-on-one personal advice. This takes the guess work out of buying a wine that you will love.'

The website will compete against major supermarket chains, off-licences and a slew of online wine retailers, including MadAboutWine, ChateauOnline, Berry Bros and WineAndCo.com not to mention BT with ItsWine.com.

Richard Branson's Virgin Group moved into the wine market earlier this year by acquiring 50% of Orgasmic Wines.

04 July 2000

Internet will usher in the euro

Like it or not, the Internet will demand that sterling gives way to the euro as Net currency says report.

One billion euros will be traded online within two years – irrespective of the British government's decision to sign up or not for a single currency, says a new report from consultants E-Insight. International websites offering everything from CDs to holidays will flood the Internet, offering the same deals to countries throughout the EU. But it will be easier for them to mark down one price rather than try to keep tabs on rapidly changing exchange rates, so the common currency will save them time and effort.

Online users will pay in euros on single-currency sites, says the report, and the amount with its sterling equivalent will appear on monthly credit-card statements – as it does now for any foreign purchase.

It also means many more Britons will be keeping an eye on the fluctuating exchange rate between sterling and the euro. A spokesman said: 'Regardless of UK entry plans or delays, UK consumers are forecast to spend nearly €1 billion – in euros – on e-commerce purchases from euro-priced sites. This will rise to almost €3.5 billion in 2003, regardless of formal UK entry plans, so these British online shoppers will have already joined Emu (European monetary union) by proxy'.

Across the EU, about €85 billion will be spent by 2003, predicts the report, with Germans paying the most – €15 billion – in the single currency.

04 July 2000

After the Net café, comes the Net pub

A bold move to develop cyberpubs is being launched by former Daily Mirror boss David Montgomerie.

Trials are underway in 20 London pubs to allow drinkers access to the Net in their locals. The attraction will be an online newspaper with funding coming from advertising. The main thrust will be to capture email addresses from people who do not have computers at home. Other attractions could include online betting.

The venture, named Yava, is looking for £25 million for a joint venture with Punch Taverns, and Carphone Warehouse is expected to be one of the shareholders. The plan is to roll out kiosks in all 4,500 Punch taverns and other targets include Bass and Whitbread which could create an estate of 14,000 outlets. The editor will be former *Daily Mirror* editor David Banks.

10 July 2000

Tesco expands into America

Tesco.com is to invest $18 million in a partnership to create a woman's portal in America.

The surprise announcement marks still further the way that Tesco [http://www.Tesco.com] has given free rein to its Internet arm even to the extent of taking virtual investments with only a marginal relation to its core business. The project is for a 50:50 partnership with iVillage.co.uk [http://www.iVillage.co.uk].

The portal will have eight channels and the investment will be spread across three years. The launch is scheduled for the end of this year.

20 July 2000

Lastorders.com sups up £2.5 million investment

The online off-licence Lastorders.com has overturned negative market speculation and secured new funding from a team of blue chip investors.

Among the investors is the Royal Bank of Scotland. 'It sends a very positive message to the market,' said former broker and head of the company James Oliver. 'One thing that appealed to investors was that we had spent only £70,000 on marketing which is nothing compared with the sums spent by other dotcoms.'

Lastorders.com [http://www.Lastorders.com] has deals with all the major brewers to deliver beers to the customer's door within 24 hours. It has a revenue target of £70 million by 2002.

20 July 2000

. . . and Eggs is looking for a £10 million benefit

The restaurant ordering B2B service from Eggsbenefit.co.uk is looking to raise further funding on the back of its signing up of a reputed 190 clients.

The facility allows chefs to order online in the middle of the night – in the same way that retailers have been doing but with the individual demands of chefs being much greater, much of the industry is still working on clipboards and phone calls. One of the prime funders is Mogens Tholstrup who sold out his restaurant's interests in Daphne's The Collection and Pasha to the Belgo group last year and has watched his profits dwindle as trading has fallen away.

Eggsbenefit [http://www.Eggsbenefit.co.uk] is claiming to have switched on 190 clients, although some of those are free but in return have created an infrastructure of introductions to suppliers. One client revealed that the software had been upgraded three times in as many months, but commented favourably that at least the service had been kept up and running.

20 July 2000

And now its 10 million, official

One in four households now access the Internet and surf for five hours a month.

The fastest growing areas are sports, hobbies, business-to-business and technology according to research from MMXI Europe. The most popular sites are Freeserve [http://www.Freeserve.com] and the BBC [http://www.BBC.co.uk]. AOL [http://www.AOL.com] sites were the stickiest, however, with users spending an average of 4 hours per month on these sites. The ICQ [http://www.ICQ.com] chat service, which is most popular with children, was the second stickiest, with users spending 2.5 hours per month using this chat tool.

The figures put Britain ahead of the rest of Europe with 4 million online in France, 9 million in Germany, and 4 million in Sweden. But the UK still lags predictably behind the 75.7 million in America, 16.4 in Japan and 11.9 in Canada. However, other research from Continental suggested the UK figure was reaching much higher – towards 14 million – which may be a reflection of current trends.

20 July 2000

Secure ordering triples orders

Cook site keeps bubbling away.

The ScottSargeant.com [http://www.ScottSargeant.com] cookshop website was set up in September 1997. Richard Britton said: 'It has always been a popular one for keen cooks and enthusiast chefs. Remarkably, it has also generated a steady flow of sales since its inception even though the initial e-commerce solution was fairly 'rustic' – e-mails, phone and fax.

He continued 'The online ordering system and secure certificate went live at the beginning of this year and orders subsequently tripled overnight. Several factors contribute to the success of our website – the products we have on offer, the fact that we are an established bricks and mortar company and our customer service.'

'One aspect which is frustrating for me is that I know we have areas of the e-commerce solution to develop but the website normally takes third place priority behind our substantial retail store and our mail order catalogue. For example, we are trying to expand our recipe book so that it is more comprehensive and interactive. I am now developing a dedicated team which will support our internet business so that changes and updates to our site will happen more frequently and in a more structured way.'

20 July 2000

Freshfood.co.uk expands organic range to 5000 lines

Pioneering web trader Thoby Young has dramatically developed his range of produce to almost 5000 lines and was recently a finalist in the Sunday Telegraph website of the year competition.

Calling it OrganicNation [http://www.freshfood.co.uk/perl5/ffc_sec/S.cgi], Young refers to the service as the world's first online nationwide grocery home-delivery service, which in practice it is though others are now racing to catch him up.

He said: 'We have recently expanded our range of organic and wild-harvested foods and related services from around 4000 to around 5500, in seven major categories and over 500 brands and sub-categories. This range is greater than the combined range of all the major UK supermarkets put together and also greater than the combined ranges of our major independent online competitors.

'In February Fiona Webster in the Daily Mail declared that the Freshfood.co.uk [http://www.freshfood.co.uk/] was no more expensive than the supermarkets on a like-for-like shopping basket of necessities. Our reliability was better, with fewer out-of-stocks, and our website was quicker to shop.

Young argues that only by developing its own in-house IT solution has the company been able to expand so rapidly. 'The site was launched in early 1996, and remarkably it still retains the basic corporate style of that original website,' he said.

20 July 2000

Organic butchers put their beef on the Net

Site lists butchers and mail order specialists for organic world.

A new site detailing names and addresses of organic butchers and mail order specialists was set up as a hobby by Frankie Button. Organicbutchers.co.uk [http://www.organicbutchers.co.uk] lists 170 outlets. Organic vegetables were getting all the publicity and he wanted to find a site for buying high quality meat.

'We have always been interested in the quality of food more than its price. The majority of the British public would, it seems, rather have cheap food with quality being secondary. However I think that there is a growing number of people who are prepared to pay a premium for quality which organic produce provides.

'From our own experience we found it difficult to track down organic meat and saw the site as a helpful resource for people looking for organic meat. I also wanted to develop a website just out of interest.

'Many of the outlets on the site are small enterprises and do not have their own sites. This enables them to have a web presence. Currently we are not charging outlets, but if the site is successful and worth continuing we will be asking the outlets to pay an annual fee, which when compared to having their own website will be very economic.

'There is a growing demand for organic meat and many more farms are switching to organic. I therefore see the site as expanding over the years. The site has been up and running for about 2 months. Getting publicity and feedback is the greatest problem, particularly as we are in effect operating on a shoestring. Our stats feedback is not updating at present so we are very much in the dark as to how the site is going. We have designed the site to operate as quickly as possible, searching does not take more than three clicks. From a personal point of view I find many sites are frustratingly slow as the designers tend to focus on showing off their design expertise rather than making the site functional. Whether the site does eventually make money is really a bonus, we have enjoyed the experience of setting up the site . . .'

20 July 2000

Gourmet gift market flourishes

American research has shown the dynamic growth of gourmet foods sold directly off the Net.

ActiveMedia has confirmed speculation that revenue generated by sites specialising in items such as gourmet food, personal care, and branded consumer products is rising steadily. Sales in 1999 are expected to have increased tenfold over 1998, while sales in 2000 are expected to be four times greater. Many of the sites are very small businesses. ActivMedia estimates that 51% of these sites will invest between $1,000 and $10,000 in their site this year, with 12.5% planning on-site investment of over $100,000.

The study found that two in five visitors to specialty stores are new users, which ActivMedia believes dispels the myth that the market for speciality goods would wane with an increasingly 'average' profile of the Internet user.

20 July 2000

Fish shop catches the big PR

A tiny fish shop in Cornwall has captured the imagination of the media, picking up mentions on TV and in the Sunday Times.

The family run business Martins-SeaFresh.co.uk [http://www.martins-seafresh.co.uk] has managed to garner the kind of exposure that other dotcoms have spent millions on. For nothing.

'We have been sending boxes of fish to various people filming on Internet shopping projects. How many will be shown on TV remains to be seen, but the interest is building. The Sunday Times did another article in their recipe section with our site getting another mention, though unfortunately they got the website address wrong and called us Martins-seafare. They did put our phone number in so it wasn't a complete loss.

'The other main change is that we have just been accepted by the *Which Web Trader* organisation and we will shortly be putting their Logo on our site, this should lend a further air of respectability to the business.'

20 July 2000

Coffee site is percolating

Specialist ground-to-order site expands by the month.

www.infusionscoffee.co.uk [http://www.infusionscoffee.co.uk] is a specialist in the 'Specialty Coffee Market'. Director Adam Badger says: 'We are offering something the supermarkets never can, we pack to the customer's requirements. Any grind, blend or origin, with over 48 coffees now in stock, we endeavour to offer a next-day delivery. Straight from the roaster to the customer!'

'They place an order, we then grind to their requirements, seal and then post 1st class. Our packaging has a unique, one way valve and zip-up seal. We supply mainly pure arabica origin coffees so the customer knows exactly what they are getting. 100% Jamaican Blue Mountain, Kasmir Cherry or our popular Kenyan Elephant – we take you around the world and you don't even need a passport!'

'Joining our coffee club gives you 10% discount, a quarterly news letter plus a 3-pint pack of the latest coffee we can find. We are always on the hunt for the strangest and highest quality coffee we can find, the last won awards for the best coffee in Brazil! A free Bodum cafetiere is also included so at only £9.99 this is a bargain. If you like coffee or know someone who does this makes a great present that lasts all year!'

'Our site has been on the net for over 3 years but was only updated in January 2000. We will have a secure server option in July, our site just gets bigger and bigger. From August we will have a complete listing of all our cappuccino accessories and machines bringing the coffee shop quality to the customer's own home!'

20 July 2000

Chicken coops go international

A specialist maker of free range chicken houses has developed a booming export trade through the Net.

Managing director of www.freerange.org [http://www.freerange.org] Spencer Lane said: 'We manufacture mainly free range poultry houses for the UK and

European market. Since the website has been in place, designed by us, there has been much more interest globally – China, Peru, USA – lots – Australia'.

'The site is well used for a commercial site – 8000 or so hits in four months – without any real push through web advertising. I would like to see our site being used as a global discussion area for many topics, and this is coming slowly. We have sold many thousands of pounds of our products off the site, but we don't take money over the Net as our products are – we think – too costly for security reasons. We find people get the info, then email or contact us and we do the transaction off line. Generally people want to discuss their purchase or requirements'.

14 August 2000

Expedia sets up Internet restaurants at airport

Travel portal Expedia has moved into high-class restaurant and deli Net cafes at airports.

At the first trial site at San Jose California, travellers will be able to book a real sandwich made with all beef kosher salami from Pearl of Boston; or, hot and spicy rock shrimp in garlic butter on mashed potatoes. Expedia.com Cafe is an airport cafe that provides a place 'for travelers to log-on and fuel up'. The first tie in is with Max's Opera Cafe, a well-known local restaurant.

Expedia has entered into an exclusive contract with HMSHost, a leading concessions operator, to have the company operate Expedia.com airport Internet cafes. 'We at Expedia have heard customers' complaints about searching an airport terminal for a place to plug in their computer and access the Internet,' said Expedia President and CEO Richard Barton. 'The introduction of Expedia.com Cafe highlights Expedia's continued commitment to helping travellers travel right by combining a comfortable, convenient place to plug in and log-on with a full-service airport cafe. We hope that this will be the first of many Expedia.com Cafes in airports across the country.'

The branded cafe joins other offline Expedia endeavours, such as Expedia Radio, a national radio show broadcast to 2.5 million listeners, and Expedia Travels, a travel magazine scheduled for launch this Fall.

14 August 2000

Hacker hoaxes Safeway

Customers told prices to rise by 25%.

A rogue attack on Safeway e-commerce security has seen more than 1000 customers receive an apparently official company email telling them to shop elsewhere and warning that prices could go up 25%. An embarrassed spokesman said: 'The message does seem to have been sent out by our own computer system'.

Surprised customers received this notification from info@safeway.co.uk: 'We are pleased to announce from Monday Safeway will be increasing prices on all our goods by 25%. If this doesn't sound good to you then you can p*** off to another supermarket chain such as Tesco or Sainsbury.'

14 August 2000

Online diet site gets fat on Nestle deal

Nestle has bought into Nutrisystem.com with a deal that gives the new company control of its Sweet Success diet replacement brand.

The shares represent a 3% Nestlé ownership stake in Nutrisystem.com [http://www.Nutrisystem.com].

'The diet meal replacement category grew at a 31% annual rate from 1996 to 1999 and Sweet Success® is a well recognized and trusted brand in that market,' said Brian Haveson, Chief Executive Officer of Nutrisystem.com. Sweet Success is an excellent complement to the nutrisystem.com online weight-loss program. We will continue to distribute the Sweet Success product solely through mass market retailers, but will create the kind of online support, interaction and information for Sweet Success customers that have proven so successful for the members of the nutrisystem.com community. We will give Sweet Success an online presence, and Sweet Success gives the Nutri System brand a presence in the 'bricks' retailing environment.'

The brand includes a ready-to-drink weight-loss shake, powdered mix and snack bars and generated over $14 million in sales in the first half of 2000.

Nutrisystem.com, which launched its online community in October of 1999, reported sales of $11 million in the first half of 2000.

Nutrisystem.com has developed the leading online weight-loss community. It provides members with a weight management program that incorporates personal online counselling, individual diet and exercise plans, a weekly newsletter and online support groups, chat rooms and bulletin boards. There are no membership fees and portion-controlled meals are delivered directly to the member's home or office.

01 September 2000

Management buys out just-sites

The information wing of the Geac software site has been bought out by management backed by 3is investment.

The Evesham-based company plans to develop the just-food.com [http://www.just-food.com] and just-drinks.com [http://www.just-drinks.com] information areas as stand-alone businesses.

The sites are currently getting 17,000 visits a week mainly for its global news feed brought in via Reuters but will now develop a food information store for specialist reports, books and market data. Director Richard Jackson said: 'Sale of knowledge is our key model'.

01 September 2000

Tesco to win half of e-grocery market says survey

A report from E-Insight, an e-commerce consultancy, predicts total sales of £75 billion by 2008 and predicts Tesco will have a share in excess of £30 billion a year.

The bad news is that by then, the company will have invested around £400 million to achieve market dominance. E-Insight is predicting that Sainsbury will invest £160 million and forecasts profits of £230 million by 2008. The high cost of entry will deter future players says the report. It says a new etailer would need £100 million of investment. The key trend will be a brutal price-cutting programme as stores reduce the cost of land-based investments and transfer savings to the Net.

In a more conservative forecast, Verdict, the retail consultancy, is currently predicting home shopping to be worth only £25 billion by 2005 with interactive TV being an important driver of growth.

05 September 2000

Campbell Soups ties in with Talk City for recipe exchange

The Campbellsoup.com site will feature recipes using the company's soups in cooking.

'For more than 130 years, Campbell has focused on building strong relationships with our customers, and we're extending that tradition online,' said Nora Schultz, Global Manager – Interactive Marketing of the Campbell Soup Company. 'Through the Recipe Exchange, Talk City helps us give our customers real, valuable reasons to visit our website. This effort should help us deepen our relationship with consumers while spreading the word about our great soups.'

Talk City is an online marketing services company providing live events, integrated community solutions and market research to businesses in a number of industries. Talk City's services help marketers to extend brand equity through the Internet and establish high value e-relationships with their customers. Partners include Alta Vista, Compaq Computer, Cox Interactive Media, Freedom Communications, The Hearst Corporation, Lifetime Television, The National Broadcasting Company, Microsoft's MSN and Microsoft's WebTV Network, and Sony Music.

05 September 2000

Real time figures on online shopping

The latest research shows women becoming the dominant buyers on the Net across all e-tailers.

PC Data Online research on Internet shoppers this Summer has shown that women aged 35 to 44 were a 'dominant e-tail force', said John Megahed, director of online research and analysis for the Reston, Virginia-based firm. 'The demographics in online shopping are shifting away from the MTV generation,' he said. The research shows the number of purchases made in a top 20 format. JCPenney.com [http://www.JCPenney.com] jumped three spots in the rankings to the No.7 position, with 218,000 unique buyers during the month. Sears.com [http://www.Sears.com] also leapt three spots, reaching No.8 with 207,000 unique buyers. Bestbuy.com [http://www.Bestbuy.com] entered the top 20 at No.19, with 75,000 unique buyers, while Staples.com [http://www.Staples.com] edged onto the list at No.20 with 74,000.

All but Staples.com saw predominantly female buyers, the report said. Other reports confirm the trend. Media Metrix and Jupiter Communications found that women now make up 50.4 percent of Web users. Earlier this year, Forrester Research said that 45 percent of all online shoppers are female.

Of the top 10 sites tracked by PC Data, only Amazon.com [http://www.Amazon.com], Ticketmaster.com [http://www.Ticketmaster.com] (Nos.1 and 2, respectively) and No.6 Real.com [http://www.Real.com] saw more unique buyers in July than in June.

Amazon.com saw a 22 percent increase to 1.8 million unique buyers, with Ticketmaster.com registering a nine percent gain to 676,000. Real.com, which offers streaming media on its site, saw the largest jump inside the July rankings, up 10 slots from June. Discount e-tailer Half.com jumped from No. 21 in June to No. 13 in July. The site was recently acquired by online auctioneer eBay.

Five companies from June's Top 20 dropped out: eToys.com, 1-800-Flowers.com, Dell.com, MotherNature.com and More.com. More.com, a seller of health and beauty products, had been No.9 in the previous month.

PC Data Online tracks unique visitors to its Top 20 sites, through software based on buying activity and traffic from more than 120,000 US home Internet users.

05 September 2000

Whoops! Not that price for WalMart!

New Walmart.com chief executive Jeanne Jackson has made surprising discovery on taking up her job.

The company site was selling goods at less than it cost to process the transaction. Needless to say her first move was to slash the portfolio dropping lines like 99c lipsticks which did not fit the e-commerce platform.

05 September 2000

One hour delivery wars

Aspirant companies looking to capture the London home delivery market are going head to head.

A third player has lunged into the market to deliver all and anything to Londoners stranded in their flats and offices. Koobuycity.com [http://www.Koobuycity.com], which looks remarkably similar to rival Urbanfetch.co.uk [http://www.Urbanfetch.co.uk], has stretched the opening hours to 1am, two hours later than its rivals. The selection of books, CDs, gifts and snack food is remarkably similar across all three contenders, the other one being Bagsoftime.com [http://www.Bagsoftime.com].

11 September 2000

Electrical store trails kitchenware

An online electrical store which fronts up its white goods in terms of DVD players, has also been trialing espresso machines and kitchens.

Slip down to the bargain basement at Bedirect.co.uk [http://www.Bedirect.co.uk] and there are Gaggia espresso machines for sale and also Siemens kitchen equipment. The name actually stands for Buy Electrical Direct, though there must be a catchier way of saying that.

11 September 2000

Europe to overtake USA in 2 years

But numbers won't matter as much as time spent surfing on the Web, where Americans will win out by three times, says Net guru.

A conference in Monaco warned that European companies will spend less on technology and fall behind in the Net race. This is an abridged summary of the key points:

Whereas the United States should be expected to continue to lead in hardware and most software, many American companies were shying away from localization and were distracted by three different mobile phone standards. European technology companies, on the other hand, were leading in mobile communications and in understanding different cultures. Instead of trying to bring the Internet to a tiny mobile phone screen, most people have not fully begun to appreciate the possibilities of computer voice telephony, delegates were told. 'This will become a tremendously important technology where people can ask for information that is then given to them automatically by a computer, either in voice or a small text,' Pim Bilderbeek of IDC marketing said.

Not surprisingly, a European company from Belgium, a country of a little over 10 million people where three languages are spoken, Lernout & Hauspie, is one of the world's leaders in language technology. Security and encryption of voice was another field that was not yet fully explored, Bilderbeek said.

Indeed, Ernesto Schmitt, co-founder of Britain-based PeopleSound.com [http://www.PeopleSound.com], said he had found his company had an edge over US rivals which were not interested in creating local sites. His company, on the other hand, was forced to address audiences in different countries. This localization enabled PeopleSound to benefit from local market opportunities, as around half of music acts tend to reach only a national or regional audience.

Gary Hamel, professor at Harvard Business School, said it could even be harmful for European companies to try and compete with US companies in technology and Internet investments. His research had found that operating margins of US companies on average had not risen over the past 6 years, despite a hefty increase in IT spending. 'This suggests that all the savings these companies have made through their investments have ended up in two places: the pockets of consumers and the stock market valuation of technology companies such as Microsoft, Cisco and Intel (which sold them the technology),' Hamel said.

He compared the Internet with the advent of electricity in the late 1900s, which had the same impact.

With the Internet becoming an off-the shelf commodity like electricity, the current 'IT arms race' would bring companies little competitive advantage, unless they added their own innovation to it. In this respect, Europe still had a lot to learn from Silicon Valley where ideas, people and capital exchanged much more freely, Hamel said.

12 September 2000

SimplyOrganic gets £1 million

Organic retail site SimplyOrganic has received a cash injection of £1 million from its original investors to bolster its home delivery infrastructure.

Ian Smith, founder and managing director said: 'The trick with B2B is to find an area of the market where the established players either can't afford to move or simply don't want to. For the supermarkets there is little to be gained, at present, from moving into the organic market in a big way. However, in the current climate, the consumer is demanding more and more choice in the organic field.'

12 September 2000

The beerhunter is bagged

Wine portal licks its lips at real ale's exclusive deals.

Michael Jackson's realbeertour.com was the froth on the deal that brought one of America's biggest and oldest wine sites Wine.com [http://www.Wine.com] to take on the distribution of one of America's biggest beer sites. Wine.com will distribute the niche exclusive beers from Realbeer.com [http://www.Realbeer.com]. Robert Imeson, president of Realbeer.com said: 'This agreement will allow us to focus our efforts on sourcing the world's most exotic beers for members of our exclusive club, The Michael Jackson Real Beer Tour. It will also provide us with the opportunity to continue to develop a vibrant and exciting e-store which will be stocked with some of the world's best, yet hardest to find, beers.'

Realbeer.com was founded in 1994 and quickly became the leading beer content site on the Web. Realbeer.com currently boasts over 125,000 pages of content and has agreements with most of the world's top beer writers designating the company as the exclusive publisher of their notes and articles on the Web. Earlier this year, Realbeer.com merged with WorldBeerTour.com, which previously partnered with Michael Jackson, the world's foremost beer authority. Under Jackson's guidance, the integrated company has been able to go directly to some of the world's finest breweries in order to provide members with great beers previously unavailable in the USA. As a result of these efforts, Realbeer.com has succeeded in developing the world's largest Internet-based beer club. Realbeer.com has plans to extend its e-commerce activities to Europe, Australia and New Zealand in the coming year. Realbeer.com is headquartered in San Francisco and also has offices in London, Seattle and Auckland.

Founded in 1994, wine.com was the first wine merchant on the Web and is the leading online wine destination. Wine.com serves wine consumers in 90% of the United States market.

21 September 2000

French make a swift exit from the crowded online wine market

The early year ambitions of French-chateau-backed wineandco.com have faded as the company has decided to pull out of the UK market.

The site had been planned to develop a full B2B portal and has opened putative offices in six countries. But has closed its London operations and retreated to France with a minimum B2C portfolio to be maintained from there.

21 September 2000

Asda parent puts all its global procurement online

In a move that will send shock waves through the retail community, Asda's parent company Wal-Mart is to put out its global sourcing via the Net.

Wal-Mart has hired Atlas Commerce to develop a global online intranet. The system will allow the company to hold online auctions for the best prices from manufacturers around the world. The move is the third major play for supermarkets following Tesco and Marks and Spencer's involvement with Worldwide Retail Exchange and Carrefour's involvement in GlobalNetExchange. Sainsbury has also invested more than £15 million in its intranet sourcing strategy.

17 October 2000

London delivery services collapse

Friday the 13th was a grim day for Internet hopes for one-hour deliveries in London with two players shutting up shop and the third going offline.

The American-backed Urbanfetch.com [http://www.Urbanfetch.com] pulled out this week citing poor market conditions. It plans to focus on its business-to-business delivery service in New York. The company had promised to deliver within an hour in the capital. 100 jobs are lost.

'We stepped back and looked at our revenues and, market conditions being what they are, we regretfully decided to end the service in London,' an Urbanfetch spokeswoman said.

Analysts have long questioned how the service would make money while heavy overheads included a large staff and fleet of delivery vans. Profit margins were thought to have been minimal and average spend around £25. Urbanfetch raised $70m from US venture capital groups. Merger talks with New York competitor Kozmo recently collapsed after the two firms failed to agree on a business-to-consumer strategy for the enlarged group.

On the same day, a second service www.bagsoftime.com also called time on its service. Chief executive Bob Hitching blamed 'apathy for e-shopping'. Theoretically that leaves the capital only with the French-backed Koobuycity.com [http://www.Koobuycity.com] which is operating in Paris and Frankfurt, but the service was also ominously 'offline' this week.

18 October 2000

Interview: Fledgling gourmet site takes wing

Promising to bring the farm produce straight to your door, a small St Albans company has opened for business boasting the legendary Neal's Yard cheeses as its first big coup.

Farmconnection.co.uk [http://www.Farmconnection.co.uk] is an ordering service that goes through to the farm or supplier to sell on their products. Director Anne Sampson's first two clients are producers of specialist cheeses and handmade preserves.

- Why did you start this business?

We realised from personal experience that unless you live in West London or have a really good delicatessen situated in your town, it's actually quite difficult to buy premium quality food. Even where we live, in St Albans, there is a surprising lack of really good food shops. There isn't even one delicatessen in the whole city.

- Surely there are other specialist shops about?

It's true there are a number of mail order 'deli' companies – most of whom seem to primarily stock dry or tinned goods, such as olive oil and dried pasta. These are ranges that the supermarkets can handle and actually do reasonably well.

In addition, there are specialist mail order companies selling really great products – but what a pain having to ring up one supplier for my poultry, one for bacon and another for oysters; each time having to fill out delivery details, write and post a cheque, or fill in lengthy credit card details. So we thought we would bring together a community of specialist producers, whose products were available to order either from the Farm Connection website, by phone or by post.

- How do you choose your products?

To make sure the products we have found reach the customer in the very best condition, we don't hold any of them in storage. Only the very best are selected by the producer and sent directly to the customer. We've had great fun setting up the website and spending time working with each supplier. We've been visiting cheese makers in the West Country with Neal's Yard Dairy, to collect handsome Cheddars, creamy Duckett's Caerphilly and see the magnificent Tymsboro' goat's milk pyramids being made and rolling up our sleeves at Rosebud Preserves to peel and core Bramley apples by hand, to make the wonderful Old Yorkshire Chutney. Really getting to know each supplier and their products. This is what makes it worthwhile.'

18 October 2000

Waitrose cuts £35 million deal to win the last mile

Undaunted by the failure of London delivery services, Waitrose has backed up its pledge to deliver down to the last mile.

Three John Lewis directors are to join the board of Last Mile Solutions, the blue-chip delivery and logistic company that has been taking a year to develop its strategy. This backs up the company's promise to have a full Waitrose product range in a new home delivery venture designed to capture a significant share of the e-grocery market.

The Deputy Chairman and Finance Director of John Lewis, David Young, the Director of Buying at Waitrose, Steven Esom, and the Head of Business Development, Charlie Mayfield, will all take board seats with immediate effect.

CEO of LM Solutions, Tim Steiner commented: 'We are now in a very strong position to redefine the home grocery market through combining cutting edge technology, retail sector expertise and one of the most dominant brands in the UK. We will establish a new benchmark which others will find difficult to match.'

This is the first time that the John Lewis Partnership has embarked on a joint venture of this scale by taking a substantial minority investment in a private independent company.

David Young, Deputy Chairman and Finance Director said 'Our investment in LM Solutions underlines our commitment to make Waitrose a pre-eminent player in e-grocery. It will provide a fast-track opportunity to extend Waitrose's reach.'

Last Mile Solutions was founded in January 1999 by three ex-Goldman Sachs bankers: Jonathan Fairman, Jason Gissing and Tim Steiner. It has already acquired some major skill sets from the retail industry including the former heads of the M&S Food Business Nigel Robertson and Roger Whiteside; Jackie Edwards, a former Director of IT at Cap Gemini; Robert Gorrie, a senior logistics professional; and Neill Abrams, a former UK barrister and US Attorney at Law. There are currently two existing non-executive directors: the former Head of Food Technology and External Affairs at M&S, Dr Tom Clayton, and Jez Frampton, Managing Director of Interbrand.

18 October 2000

Tesco pips Foodferry.com for top rating

A new tracking method on Internet sales and service has rated Tesco number one.

Gomez, the Internet quality measurement firm, examined areas such as customer adoption and successful integration of click-and-mortar e-tailing strategies.

Of the six online supermarket firms that met Gomez's minimum criteria, Tesco topped the scorecard followed by Sainsbury's and Foodferry.com [http://www.Foodferry.com]. The highest rated sites offered the most robust online capabilities, widest product selection, and home delivery.

'Many large UK supermarkets have turned to the Internet to deliver more value to their existing customer base and to ring up incremental sales from Internet-hungry shoppers whose current grocer doesn't offer an online solution', says Matt Stamski, senior retail analyst for Gomez. 'While there are only six players on this Scorecard, the competition is fierce, as is proven by the sophisticated efforts of the supermarkets compared to other e-tailing channels.'

The Scorecard examines each site through the same lens and breaks down the online experience into distinct categories, such as Ease of Use, Relationship

Services and Onsite Resources. Shoppers and professionals can sort the rankings based upon these categories to help them find the sites best equipped to suit their needs.

Tesco topped the scorecard as Best Overall Supermarket, Customer Confidence and Onsite Resources. Foodferry.com was rated as the best site for Ease of Use and Sainsbury's was nominated as the site that provided comprehensive Relationship Services.

Gomez is a provider of e-commerce customer experience measurement, benchmarking, and customer acquisition services.

18 October 2000

Spice giant predicts Web will boost profits by five per cent

McCormick plans to enable the supply chain through online trading.

'We will begin in earnest by early next year', vice-president and treasurer Christopher Kurtzman said. 'Our Beyond 2000 system and process initiative will create the ability to further improve efficiencies and to return gross profit margins to historic highs approaching 40 percent,' he said, which would be nearly 15 percent above current margins of 35 percent.

McCormick dominates its niche. On a revenue basis, the company claims a 36 percent share of the spice and seasoning market. On a volume basis, it sells 22 percent of all spices, according to Merrill Lynch. Kurtzman said: 'We continue to make progress in improving cost information, which allows us to segment customers and take appropriate pricing action.

Prior to last week, McCormick had kept mum about e-business, and company officials declined to elaborate afterward. 'At the front end of the supply chain, we are combining material requirements across international operations to gain leverage with, and reduce the number of, suppliers,' Kurtzman said.

A big part of McCormick's Beyond 2000 strategy involves B2B software from Logility Inc., which was enlisted by McCormick to implement Logility's Voyager Solution commerce package.

McCormick licensed demand planning, inventory planning and distribution/supply planning applications aimed at 'enabling McCormick to forecast what customers buy from their sales reps,' said Andrew White, Logility's product strategy vice-president. 'It's a two-phased process to sort out internal supply chain issues and optimize internal operations, then implement demand planning and inventory planning tools.'

Voyager integrates with existing ERP apps through prepackaged interfaces to SAP and Oracle.

18 October 2000

Iceland taking £600,000 a week online

Internet orders are worth six times the average store spend.

Iceland has revealed plans to invest £2.4m in a data-warehousing project. In March, the company reiterated its commitment to the Internet as a selling tool, changing its name to Iceland.co.uk. IS director Martin Chatwin revealed: 'Internet ordering is already equivalent to four or five new stores at 10% of the cost.'

Each of the 10,000 online orders received every week are worth an average of £60: 'Six times that in store and 20% higher than orders placed through our call centres', said Chatwin.

With the implementation of new Teradata-based Worldmark servers, Iceland hopes to further increase its online grocery market share. Chatwin explained: 'We wanted one version of the truth across the whole group, and to get a single view of what customers are buying across all our channels.'

20 October 2000

Net diet is endorsed

One of the most successful Web diet sites has had its programme validated by an obesity research centre.

Obese postmenopausal women who used the Nutri/System Meal Plan for 16 weeks lost approximately 21 pounds or 10.7 percent of their body weight according to a recently published clinical study. 'The risk of developing chronic diseases with aging is a major concern for postmenopausal women; however, this study shows that the elderly can achieve a healthy weight loss using portion control with a nutritionally sound diet,' said Steven B. Heymsfield MD, a deputy director of the Obesity Research Center at St. Luke's/Roosevelt Hospital and a co-author of the study.

Nutri/System combines the well-established NutriSystem name and weight loss program with the Internet as a medium of communication. Through the website, www.nutrisystem.com [http://www.nutrisystem.com], the company provides a comprehensive weight and lifestyle management program that includes real-time personal counselling, individual diet and exercise plans, portion-controlled meals, online support groups, chat rooms and bulletin boards.

20 October 2000

Tea sales shoot up 450% on back of Java

Celestial seasonings website reported spectacular growth with the introduction of Java.

'We wanted to exceed our customers' expectations when they shop celestialseasonings.com [http://www.celestialseasonings.com],' said Steve McKown, IT director for Celestial Seasonings. 'We needed the flexibility to improve the site's e-commerce functionality for customers, in particular the integration with our fulfillment systems. At the same time, we had to implement the new site quickly and at a reasonable cost. We saw ATG's Dynamo suite as the most balanced offering available in the market and the best suited to address our requirements. Since improving the functionality of the site, our visit-to-order ratios have improved dramatically.'

The largest manufacturer and marketer of specialty teas in the US, Celestial Seasonings' direct business unit markets tea and products that extend the tea experience directly to end consumers. The company launched their original retail store, located at the factory, in 1991, followed by the introduction of the gift catalogue in 1994. In 1997, the website was launched. Consumers visiting celestialseasonings.com can purchase teapots, honey, travel mugs, herbal gift sets and branded merchandise like Sleepytime(r) nightshirts in addition to tea and supplement products.

20 October 2000

Hic! It's a virtual bar

Talk to yourself at the Heineken chat lounge.

Lager brand Heineken has set up a virtual bar including chat to promote their beers worldwide. Heineken employed IBM Pervasive Computing consultancy and professional services to construct BarTrek.

'The world of Heineken is a world of people – and no other market is as diverse as the market of world beer drinkers,' said Jose Evers, Heineken's Head of Interactive Marketing. The solution design required Heineken to select bars within 15 cities, and have them photographed and described down to an exact GPS reference. These bars were then marked on digital city maps on Heineken's BarTrek website. Once BarTrek users have downloaded the BarTrek application from the site to their PDA device, each individual map is available for selection and download to the PDA. When travelling, users are then able to connect their PDA to a GPS unit which will display the user's exact location within the city (and update this every few seconds) in relation to the bar towards which he or she is heading.

Heineken's IBM-built site, claims to receive 2 million hits a week and has 100,000 registered members. But when Efoodnews.com checked it out we were the only people in.

20 October 2000

Hole! Mex goes Tex

Bilingual Spanish/English Mexican food shop gets hot online.

'MexGrocer.com [http://www.MexGrocer.com] is designed to cater to the 34 million Hispanics in the US and the ever-increasing number of non-Hispanics that crave and enjoy authentic Mexican specialty foods,' said Ignacio Hernandez, the company's founder, president and CEO.

'We have also created our own niche by offering consumers and food professionals these non-perishable, hard-to-find Mexican grocery items online, where they would normally have to look anyway.' MexGrocer.com has formed a strategic relationship with Royal Crown Foods, one of the largest Mexican food distributors for warehousing and for order fulfillment.

'As a value-added service,' added Hernandez, 'MexGrocer.com provides professional and amateur chefs with authentic Mexican recipes and cooking tips from our prestigious executive chef Patricia Quintana, who has devoted the past 30 years to cooking as a profession and has accumulated extensive experience in native Mexican cuisine.'

20 October 2000

Tesco says 35,000 grocery orders a week come off the Net

Of its 60,000 orders a week more than half are for grocery shopping, e-commerce director John Browett says.

But he has resisted the call to follow the Sainsbury and Webvan model of dedicated depots. He said: 'We are not against picking centres. It is just that you need the volumes to justify them.' He is now claiming that 90 per cent of the UK is served by www.tesco.com. He also dismissed the idea that customers are only using the service for bulk items. 'People start by saying they are going to do their bulky shopping online. But eventually they just do all their shopping', he said.

23 October 2000

Now Nestle tries its hand at recipes

Nestle has launched a new site VeryBestMeals.com dedicated to itself.

The site features recipes from its 16 major brands including Buitoni, Nestea, and Nestle Carnation Instant Breakfast. The site also incorporates personalized meal planning tools, such as My Meal Plan, My Recipe Box and My Grocery List.

'We have been working with Meals.com [http://www.Meals.com] as a client on their branded and retailer sites for some time,' said Ken Crites, Nestle E-business Marketing Manager. 'They have proven to us they have the people and technology in place to create and service a top-flight website. We are very pleased with the results and see this as a positive step in the development of our online presence.'

'This is a tremendous step in the growth of Meals.com,' said Mark Seals, COO of Meals.com. 'We are now an application service provider for the world's largest food company. By building and maintaining VeryBestMeals.com [http://www.VeryBestMeals.com] for Nestle, Meals.com is establishing its technology in the very competitive online food marketing arena.'

Although powered by Meals.com, the look and feel of the website is customized to support Nestle's brands and image. VeryBestMeals.com complements Nestle's other online ventures, including VeryBestBaking.com [http://www.VeryBestBaking.com], VeryBestBaby.com [http://www.VeryBestBaby.com], and VeryBestPet [http://www.VeryBestPet.com].

Meals.com Inc. connects consumers, grocery stores and manufacturers through on-line technologies that provide shoppers with relevant information and special offers based on their personal preferences. Meals.com is a subsidiary of Coinstar Inc. (Nasdaq: CSTR), a coin counting and e-services company that has relationships with more than 8,000 US supermarkets.

17 November 2000

Foodoo folds

Star recipe site falls flat.

The recipe site that boasted Gordon Ramsay, Sophie Grigson and Hugh Johnson among a list of top names supplying recipes and wine advice has closed after four months because it was unable to secure further funding. Since June, it claimed 12,000 had registered and the site was getting 30,000 page impressions a day.

17 November 2000

Del Monte sells fruit direct online

Claiming a world first, the man from Del Monte has said yes to selling direct to consumers.

In the first stage of a planned roll out that will include full B2B applications, Del Monte has launched www.fruits.com [http://www.fruits.com] to sell straight to consumers. Mohammad Abu-Ghazaleh, Chairman said: 'Our new strategies mean new revenue streams, new growth opportunities, improved forecasting and planning, and improved operating efficiencies. We are looking forward to the opportunities this new B2C initiative will create for Fresh Del Monte to expand our brand awareness and product loyalty.'

Consumers can buy pineapples through to complete decorative hamper baskets delivered across America from $9.99 to $89.99. For more information: www.freshdelmonte.com [http://www.freshdelmonte.com].

17 November 2000

More than 7,000 register turkey protest

A website attack on Bernard Matthews has brought more than 7000 visitors to complain about the smells at one of his farms.

The canny protesters bagged the www.bernard-matthews.com URL before the company had wised up to the Net and has used it to launch the British Action Group Against Turkey Smells. The villagers at Briston, near Holt in Norfolk, complain the smell is so bad that they cannot even let their children out to play in the back garden. The site was launched in July but has been gathering signatures and petitions as it goes ever since. The site was set up by data specialist Trevor Holmes who said: 'There are half a million birds on this farm and on a hot day the stench is truly awful'.

01 December 2000

Now Asda lays out its battle plan

After months of speculation, Asda has finally revealed its Internet strategy.

It plans to roll out home shopping to cover 60% of the UK by 2002. It expects to create 2000 new jobs a year. Two dedicated picking centres are located at Croydon and Watford and others will follow as demand picks up. Trading director Mike Coup said: 'The advantage of picking in-store is that it enables you to roll out the service very quickly. But once you have got a lot of people using the service it is better to have depots'.

He said he expected much of the ordering to come via mobile phone – especially for updating regular orders – and the shopping service on Sky TV's Open network which launches in January.

01 December 2000

Eggs benefit scrambled

Liquidators move in on start-up that reportedly spent £3.6 million in 6 months.

Eggsbenefit.com which promised to streamline ordering in restaurants has called in the liquidators with a reported £1 million in unpaid debts and a further £2.8 million lost in the process of signing up around 100 restaurants in London. Only in July the company had been looking for £10 million of further funding.

04 December 2000

FoodfromBritain.com gets a new look

Export drive is behind next generation of technology.

The new site aims to provide food and drink exporters with all the information and contacts they require at the touch of a button. The latest market statistics are flashed on the site's home page, together with a currency converter with real time exchange rates. A full list of the exhibitions and exporter events around the world organised by Food from Britain can be easily accessed, together with a catalogue of publications and research reports which can be ordered on-line. Exporters can even forward requests to Food from Britain for tailor-made research packages via the site.

Simon Waring, Marketing and International Management Director for Food from Britain said: 'Exporters are increasingly using the internet to research opportunities in international markets and to obtain relevant and up-to-date data to help them make their business decisions'.

18 December 2000

Part 3
Analysis

Reading behind the riot of ecommerce headlines can be like trying to foretell the future in the tea leaves. We look in depth at the real trends that are going to fashion and shape the world of ebusiness.

Analysis: supermarket samurais or kamikaze

The New Year has broken with two astonishing pieces of news from the supermarket arena.

Waitrose is about to launch an organic box vegetable scheme directly through one of its third party packaging suppliers. And Tesco has been canvassing suppliers asking for £10,000 for the privilege of accessing its Intranet services.

These two isolated fragments of news focus the mind on just the kind of spin that major retailers have got into with the Net. Either or both retailers would seem to be on the point of abdicating their traditional stranglehold on the market place. Tesco certainly, Waitrose perhaps, are ambitious enough to believe that these trial runs may just be part of the great Net learning curve. But the implications run much deeper.

First, the Waitrose move on boxed vegetable schemes: the side issue here is about the organic movement. Waitrose has been a consistent winner of the organic supermarket of the year awarded by the Mail on Sunday. In one sense its move to develop its organic services might be seen to be enlightened and experimental. However the small organic farmers who have pioneered so successfully the notion of home delivered vegetables – and often at very high quality and value for the customers – are hardly likely to be enthralled by the arrival of a major retailer on their patch. So far they have eked out a living by skipping the middleman and shop, and now they have competition from both. They may well be feeling that there really is no escape from the retail behemoth.

And from a customer's point of view, they are entitled to wonder whether Waitrose is going to be worth its usual 50% margin, or even a considerably lower figure.

The big question is what does Waitrose contribute in this equation? The vegetables are being sourced, packed, the money taken and the orders delivered by a third party. The Net has simply provided the marketing means. And because Waitrose has swept up a number of loyal customers who are making their first tentative shopping moves on the Net, the retail giant has simply put up an icon which is no more than an advert. Waitrose has in fact taken the first step to becoming a virtual retailer. Apart from the entry point on the Net to anyone who happens to know its brand name it is doing nothing but facilitating a direct sale. By all accounts it is currently paying for all the marketing costs and as such is performing simply as a third party mediator, or actually as an advertising agency. Take away the bricks and mortar element of the stores themselves (and the huge logistics effort behind that) and what is left? The answer is very little.

In the short term it may well secure a competitive edge in the boxed delivery market, but as it does not own the lorries, the credit card machines, the packing houses, or the farms, the commercial position it has snatched is vulnerable to attack from many sides. Take away the need for shoppers to actually visit the shop and what is left but the brand name and a web site. Millions of people are not about to change their shopping habits overnight and give up the weekly trek to the supermarket in favour of web shopping. And many too will continue to favour buying organic vegetables from small producers with convincing pedigrees like SimplyOrganic.net, or Freshfood.co.uk or Organicsdirect.co.uk.

But investors – or in this case the partnership which allows employees to benefit from the dividends created by the company – might wonder where all this is going to lead. If the company is just to be a marketing brand name then the traditional grasp retailers have had on the producers might swiftly change. Suddenly those with the logistics and skills simply to deliver product will become more powerful. This new avenue of trade is also going to eat away at conventional purchases because shoppers do not have to go to the supermarket at all.

The retailer is caught in a cleft stick. If they do not do it, then the way is open for someone else to, but if they do it then they are damaging their own core trade. From an e-commerce perspective the move is innovative and dynamic, the strange thing about it is what on earth does a major retailer think it is doing handing over its commercial responsibility so completely?

In a historic sense the retailer's market position is controlled by the simple geography of being close enough to enough shoppers to command the commercial heights. But in e-commerce that gilt-edged guarantee of customers no longer holds true. Theoretically if e-commerce goes forward at the anything like the rate that analysts are predicting, this move might be seen as the first in an on-going cycle which eventually will make Waitrose as we know it redundant. There is a fair way to go before that happens, but the writing is on the wall. And it is worth remembering that to date Waitrose has been ahead of its rivals in its thinking, especially with its Waitrose@work scheme which provides office delivery direct to companies with more than 300 terminals and which claims to be the first profit making on-line food retail project on the Net.

The second equally stunning news is from Tesco. Suppliers have been asked to pay over sums of £10,000 plus a monthly subscription to access up to the minute data on sales and delivery patterns. Those used to supermarket practice may not raise their eyebrows too high on this one. But beneath the superficial indignation, the strategic implications are hugely significant.

As has already been seen in America, the traditional source of market and supply line information is not necessarily coming from traditional publishers or other, overtly or covertly, so-called independent third parties. Monsanto has set up Farmsource.com which is ostensibly the Farmers Weekly of the future. Cheaper to have a web site resource than to advertise.

By all accounts the Tesco Intranet is an admirable resource with information as various as forecasts from sales to weather. But there are times when such resources can be too powerful. If a manufacturer can access day to day sales and statistics of how each line is selling, then the day will come when that information can be used for other commercial ventures beyond Tesco. For the time being no serious threat exists to Tesco as a major front door to the British shopper. But an Internet-delivered solution could bypass that function quite quickly if the power of home shopping increases as predicted. Already in New York Priceline.com has come to arrangements with wholesalers, manufacturers and different retailers to allow customers to price their weekly shopping baskets on the Net as they see fit – as economically as is realistic.

Suddenly the consumer is empowered not just by the convenience of home delivery but by price. And as all the evidence shows, shoppers seem to be happy to shop on-line for mundane commodity articles such as toilet rolls and washing up liquids, where quality and discernment no longer seem to retain the difference between brands.

Even a small percentage of customers choosing to shop in this way would have massive impact on store profits. Supermarkets would suddenly be deprived of the revenue from impulse purchases on which they rely for a healthy wedge of their profits. Equally, brand advertising, which has sustained commercial TV for the best part of three decades, might start to gravitate towards the Internet in order to put itself in front of this new generation of shopper.

In a dynamic volatile market place like this, giving away the market data on which any business decision might be made seems like a kamikaze act.

04 January 2000

Packers plan organic box delivery via supermarkets

In a revolutionary step forward a major packaging company is in trials with Waitrose to launch box delivery of fruit and vegetables nationwide.

Talks are also underway with Somerfield's 247 home delivery service and Sainsbury's. The Lampeter based Organicfarmfoods.co.uk [http://www.organicfarmfoods.co.uk] acts as a distribution point to package organic produce for multiple retailers. The new service with Waitrose is expected to go live in the coming week. IT manager Nick Hermann said: 'We can fulfil process and take all the orders and promise delivery within 48 hours nationwide via Interlink. Waitrose will do all the marketing on their web site'.

The move will send alarm bells ringing through both the retail and the organic movement. For retailers it marks the first step into virtual delivery of products which they have not touched, while smaller organic producers who have relied on their own box vegetable schemes for a unique source of income in recent years will be concerned that multiples are muscling in on their territory.

It is also a serious new dimension for a packaging company that can now use e-commerce to sell direct and effectively become the platform for a major high street brand. To date the Organicfarmfoods.co.uk site has been largely used to fish for new suppliers. Over two years it has introduced 15–20 new suppliers. Hermann reported 'It has been steady but hardly tangible. It has also introduced researchers and academics and other useful people, plus it gives customers a good perception about who we are'.

The company, without being large in retail terms, is one of the bigger players in the organic market with a presence in America and in South Africa. Hermann also discounts the notion that there is not enough organic produce around to meet supermarket needs. 'They are unable to take all the organic produce that we can offer, plus we are now finding that some of the more conventional companies are also getting into organic, so it is all very dynamic and exciting'.

04 January 2000

Analysis: Net GM campaign starts to bite

It's Pharmageddon as Web activists take on big food companies.

The latest target is Frito Lay corn chips. Using the Web, one company a month is being targeted until it confirms in writing that its products will not use GM foods. Heinz and Gerber have already been targets and caved in under pressure.

Orchestrated by the Organic Consumers Association [http://www.purefood.org], the monthly newsletter quoted the Wall Street Journal article stressing that companies involved in GM foods would expect to see the campaign hitting stock values.

In December the OCA site set up a 1000-strong protest outside the American Food and Drug Administration buildings in Oakland, California. More than 10,000 people were contacted as part of the campaign and a tie-in with the newly formed Gean – Genetic Engineering Action Network – saw volunteers distribute more than 20,000 leaflets against GM foods outside supermarkets.

Last month another 1000 demonstrators marched through Montreal, denouncing the governments of the so-called 'Miami Group' (the US, Canada, Australia, Argentina, Chile, and Uruguay) for trying to subvert a Biosafety Treaty that would regulate the multi-billion dollar international trade in genetically engineered foods and organisms. Behind-the-scenes activity has been no less intense. A federal lawsuit against the Monsanto corporation, alleging that Monsanto had engaged in monopolistic business practices and had begun commercialising genetically altered crops without first ensuring they were safe for consumers and the environment was filed by the Foundation on Economic Trends.

Credit Suisse First Boston, added fuel to the fire by pointing out that major food corporations are running scared and that 'if anyone is in control it appears to be environment and consumer groups'.

In a long-anticipated move, Monsanto's major stockholders forced the company in December into a planned merger with pharmaceutical giant Pharmacia & Upjohn and to spin off its controversial, debt-ridden agbiotech division into a separate company. The Wall Street Journal said the merger 'is likely not only to push biotech to the backburner, but also to cost Monsanto its independence. . .'. The move also prompted the new word Pharmagedon as a label for the section in http://www.purefood.org web site. Reuters reported that US farmers plan to 'cut back sharply' on planting genetically engineered soybeans, corn, and cotton this year, in response to the growing global backlash. Farmers told Reuters to expect reductions of 15% in RoundUp Ready soybeans, 22% for RoundUp Ready corn, 24% for Bt corn, and 26% for Bt cotton. Whole Foods and Wild Oats, the two largest natural food supermarket chains in the US, plan to ban genetically engineered ingredients from their hundreds of private-label products. The two companies have combined sales of almost two billion dollars annually. The OCA was claiming also that as a result organic foods were the fastest-growing and most profitable segment of American agriculture, with projected sales this year of $6.6 billion, representing 1.5% of all grocery retail sales in the US. Meanwhile hamburger giants Macdonalds, Wendy's and Burger King have privately admitted that they will not be buying GM potatoes, although the oil they are used to fry in is believed to be derived from GM sources. But the pro bio tech lobby has not been idle either. A multi million dollar PR campaign has been launched supporting bio technology and the Clinton administration is expected to issue new guidelines on labelling soon. Novartis has launched a new website where, among other things, you can send off for a car bumper sticker inscribed with the slogan 'We Back Biotech.'

22 February 2000

Those denial attacks – chapter and verse

The first wave of guerilla attacks on portal sites was no real surprise.

Even as Denial of Service attacks swept around the globe, disabling the sites of many of the biggest names in Internet business, experts on the Carnegie Mellon University's Computer Emergency Response Team had already issued a global warning that such attacks were imminent. The team warned that recent postings on the Net of tools to automate and co-ordinate mass e-mailings targeting particular services, indicated that a cyberwar was brewing as far back as December.

The war began with Yahoo on the Monday, and then on the Tuesday they took out Amazon, eBay, Buy.com. CNN and others. All had their servers deluged with data packets that brought them crashing down for hours at a time. The next day it was the turn first of E*Trade and then ZDNet, and finally Datek Online.

The only gainers so far have been security firms such as Keynote Systems, Axent, and ISS group, all of which saw their share prices leap. Currently all the attacks have been on sites listed in the top 25 most visited sites in the United States. The US Commerce Department asked all companies and agencies to be certain that their sites are not being used as launching pads for the attacks. Yahoo said the bogus data packets came from 50 different Internet addresses at rates of up to a gigabyte a second – an enormous amount of web traffic over such a short period.

22 February 2000

Analysis: Has the bubble burst . . .

. . . or is this just the beginning? The market place for late 2000.

Read the newspaper stories and you might believe that the IT market is in tail spin and that all the bright new Internet companies are teetering on the verge of bankruptcy.

The old newspaper adage is that you build up stories and then knock them down again. The current wave of panic about Internet stocks is typical. Of course there have been some outrageously stupid Net investments from institutions And reading between the lines, there would appear to be a cohesive campaign to warn small investors from frittering their savings away on virtual gambles. But in practice these trends simply underline the growing maturity of the Internet market.

It is alleged that many Net companies will run out of money within 15 months. The caveat is of course that they carry on spending on marketing and investing in technology at the present exponential rate, neither of which seem likely.

Technological investment comes in waves and as can be seen from the American model, once a business is up and running the pragmatic constraints of cashflow soon put the brake on new distractions. And marketing money has been used to establish the brand and the medium with gay extravagance. The current jargon in advertising circles is buying fame. But this again is a flexible budget.

In Paris last month at a meeting for entrepreneurs and venture capitalists, the latter outnumbered the former by four to three. And contrary to the popular myth in the newspapers, Internet start-up companies are not finding it that easy to raise money. The old rules apply: Who is the management? What is the idea? All the pointers suggest clearly that the door is closing on wildly imaginative new start-ups and the serious money for the next two years will divert into the e-commerce development of conventional businesses.

Whittards, the coffee merchants, has reportedly invested £800,000 on its Web approach. Looking at that sum from the perspective of a website as a supposedly free medium, it may seem like a lot of money. But in practice, such an investment represents no more than adding three or four shops in the high street. And the upside is a dynamic growth in conventional mail order and the very real opportunity to trade globally. And fresh coffee is a well-rounded Net product – high premium value, light to pack and post, no spoilage, regular usage, and in some parts of this country and elsewhere in the world, not easily accessed. Turn the equation on its head and from the high street point of view, the closure of three or four shops of a major chain is hardly presented as folly or a catastrophe. Rather it is the usual cut and thrust of corporate branding.

In California, down in Silicon Valley, we are led to believe that the launch of new Internet ventures has spawned party groupies, so regular and fantastic are the launch events. Certainly America looks overdue for some creative overdrive.

Even if the UK marketplace looks as though it is overcooking, the opportunity remains clear cut: on the doorstep of Europe, English speaking, with a proven record of creativity in the related media of TV, film and advertising. The true strategists see the opportunity for what it is – a new American market about to develop, and in the last year the UK has demonstrated that it is able to steal a march on its neighbours: Germany is hamstrung by language, France by its parochial dependence on Minitel, Italy by the lack of penetration of PC users. *Virtual Europe* is there for the taking and the UK is in pole position.

None of this excuses or denies the more extreme follies of some Net projects or the presumptions that profits will just fall out of hyperspace. The Net has always been about timing. But equally, just being virtual does not rule out conventional economics. We are told that nine out of ten Internet businesses may go bust in a year. But in the food and drink sector the number of trading companies is much smaller than is popularly believed.

Most companies these days accept the need for a Web platform but for the vast majority that still means an investment from the marketing budget. These are not businesses as such, just websites that are no more than business cards. And there are numerous wannabe businesses that have floated websites on the off chance that they make a few bob. How many of these are real businesses and how many are simply opportunist bits of fun is another question.

The real successes have been where companies have applied an Internet dimension to their existing business. Tesco.com for example has left the rival supermarkets for dead. Or at the other end of the market the Yorkshire butcher Frank Scaithe's has now closed its bricks-and-mortar butcher's shop and concentrates wholly on the Net. In both cases we are witnessing a complete transmogrification of the business model.

The true winners are companies that are involved in logistics and parcelling, or development companies that have seen the opportunities that technology can bring to the running of a business. In that arena there are huge gains still to be made as management teams slowly grasp the opportunities and learn new ways of working.

The virtual sales presentation for new products or new strategies is still in its infancy but logically, creating e-brochures that can cover every aspect of a new service and be beamed to the would-be buyer is inevitable. Faster, more efficient, cheaper, more detailed Internet brochures must have a major impact on conventional selling. Take that one stage further, as we have seen in America, and we should not just presume that conventional trading will survive. Customers may set their own prices. Auctions can be instantaneous. Loyalty may come into play with electronic rewards. The gap between prime producer and customer will certainly narrow.

Take this article itself: today, if you are a registered efoodnews user, you receive it for nothing. We have no print bills to pass on. Next week, this article will join others in the archive and will only be available to subscribers for £200. Later this year I anticipate that you may access the archive just for this article, or even the bits of this article that are relevant to you and you may pay just 3p a word, just to get exactly the information that is most valuable to you or your business. And anyone resisting the process and wanting this information in a more conventional way, will have to wait and probably pay considerably more.

The same systems will apply to conventional businesses, most obviously farming where accreditation can be done online. Due diligence and other health and safety issues can be wrapped up in one simple website and customers mandated to pay for it to demonstrate their allegiance to law. And with electronic sales, accounts will automatically process the tax and VAT issues and even allow sophisticated forecasting and new generations of consultancy. This does not make everyone in the conventional business redundant but it does refocus what they do in a much more people oriented way.

The spectacular money may well drift towards those tiny start-ups that are suddenly able to grab a significant share of market. But it will be these companies that are most at risk. They will need money and will still have to raise it in conventional ways which may or may not be easy. And consolidation seems less of an answer.

Rival Internet companies are going to have equally inflated ideas about their own value and as such the prospect for mergers looks dim. And one of the truths about the Net is that intellectual property is hard to protect. So the winners in what is increasingly looking to be a dog-eat-dog affair will be small operations with strong IT links who are able to respond swiftly, and better their competition.

Another suspect area where the sands of commerce must change is in the value of an e-mail customer list. Customers change their e-mail and will certainly do so over the coming months once the BT monopoly is eased. They will also change and upgrade their computers necessitating another potential loyalty change. The huge numbers of surfers claimed by portals such as AOL and Freeserve are by no means sure-fire customers. And similarly smaller niche lists from specialist companies will date very quickly. The exponential rate of change on the Net means that surfers are very likely to be promiscuous. Business graphs need to change to take account of this.

Very likely small operations will be gobbled up by conventional businesses looking to buy themselves specialist management and a platform, but even here the idea that the Net is free and cheap is pernicious. And the smaller players will be confronted by larger corporations trying to do it on their own. In the hotel sector Hilton, Granada and Accor are trying to bind themselves together into a virtual booking service and cut out the middlemen. Whether they will take the next step and develop this into a trading environment is not yet clear. No matter perhaps at this stage that such a site is unlikely to generate too much consumer traffic, or that having a Net presence is necessarily any guarantee that the basic business proposition remains the hotel room themselves.

In time it will be plain whether such groupings have the Net vision to make their site a successful Net venture, or if they have to support it massively with marketing money or just bail out and hand it over to those who have cut their teeth at the sharp end of the Internet.

Corporate greed demands that the shape of the Net market place will be defined in the coming months by the ability of these companies to effectively manipulate the Web market. And that the survival of the smaller players will be down to their wit and savvy. There is no guarantee that any number of MBAs is going to be a relevant qualification for virtual economics.

Herein lies one of the great fascinations of the Net. Most of those fluent in the applications of Web building and developing are still under 30. Most of those fluent in basic business building are over 30. How these generations interact remains a huge intangible. The Net is now too complex a beast for anyone just to 'plug and play' and so both groups are essentially interdependent. Major areas such as e-PR, e-marketing, e-services remain unexplored and just waiting for a skilful predator.

There remains one other sleeping behemoth which has yet to fully play its hand. The media groups. One suspects that Rupert Murdoch was not exactly rushing out to invest in LastMinute.com, and yet eyebrows were raised when he paid the best part of £250 million for an Internet sports company. Now, however, Sky TV can interact through the Web and show every goal from the last ten years of premiership football. That is a resource that has never been available before. That is the kind of new product that the Net will spawn. And the same will happen in the food and drink sector. The drill down of information, the ability to source to complete niche markets, and the chance to create new markets altogether: these will be the real seismic changes of the coming decade.

Effectively what the Net will do will be to ask companies to reconfigure who they are and what they are doing. So far we have only witnessed the first tottering steps of a brand new market. This story is going to run . . .

19 May 2000

Analysis: the rise of an Internet success story

From Sunset Boulevard in old painted Beetles, Pink Dot.com has become a feature of West Los Angeles.

One of the most romantic stories of Net successes comes predictably from America's West Coast. This is an unedited version of how it came about which has a certain literary beauty and deserves full marks for the marketing department. I make no apology for running it in full.

'It all started with the Jetsons. Pink Dot Founder, Bill Toro, was fascinated by the cartoon's depiction of a television-phone that instantly delivered whatever was requested. How could he make that happen in the real world? The future presented itself in an unlikely package: a worn-out liquor store on Sunset Boulevard was for sale.

Bill bought the store in November of 1986, remodeled, and in January of 1987, Pink Dot was born. By May, Pink Dot had started delivery service, and with it, the evolution of a state-of-the-art system. With little funds for advertising and drivers crisscrossing the streets of West Los Angeles, the Pink Dot fleet was initiated. Utilizing vintage Volkswagen Beetles, the cars were custom painted a bright blue with pink polka dots. Crowned with a propeller-topped beanie and a giant, pink wind-up key, the Pink Dot fleet had the look of oversized toys. The response was tremendous. Everyone wanted to know: What is Pink Dot? As the business grew, the concept of a closed warehouse started to take shape. Again, with an eye to the television-phone idea, Pink Dot strived to replicate the in-store shopping experience without the hassle of leaving home or office. As with any new experience, (faxing instead of mailing, using an ATM instead of interacting with a teller) some customers were hesitant to try the Pink Dot way of shopping. However, as thousands of repeat customers will attest, once they have had the experience, Pink Dot quickly becomes a way of life.

Encouraged by continued success, the Pink Dot concept was further crafted to better serve clientele. With the January 1995 launch of Pink Dot / Santa Monica, the original inventory, layout, quality control and delivery procedures had been considerably enhanced. With a winning formula, Pink Dot added locations. By tracking sales, Pink Dot maintained a consistent product selection tailored to its customers' tastes. With more items than a standard convenience store, but less than a major supermarket, Pink Dot found a winning combination of carrying brand leaders in the leading sizes and quantities. A quantum leap toward the

Jetson's model was realized with the advent of the Internet. Pink Dot saw both the challenges and the advantages of Internet access and went to work to harness this sales opportunity. On May 13, 1997, www.pinkdot.com [http://www.pinkdot.com] was launched. Internet orders have steadily increased.'

Pink Dot must be doing something right when THE Ben and Jerry of Ben & Jerry's Ice Cream travelled across the country to visit their 'sales per square foot' leader in the nation, and when the former president of Kinko's decides to head up a company that had less than ten outlets – the age of the television-phone delivery experience can't be too far away.

04 July 2000

Analysis: the story of a one man start up

Hot stuff in the pepper markets.

Malc Newton launched Pepperama.co.uk [http://www.pepperama.co.uk/] on a shoestring. He knew about computers and IT. He knew he liked peppers. He explains the thought processes of why he has gone the way he has, setting up a new business for less than £500.

'I've been toying with the idea of starting a business selling over the internet for over a year but, despite having the technical skills, I haven't had the time to build a store from scratch because my 'day job' is too demanding. There were just too many hurdles:

- How do we process credit cards?
- Who will host the site?
- How do we integrate shopping cart software?
- How do we arrange delivery?
- What Insurance issues are there?
- How much does it cost?
- Where can we get the product?

'But I still had an urge to sell hot pepper sauce on the Internet. Why hot pepper sauce?

- I like hot pepper sauce and it is not widely available in the UK.

- There are hundreds of online stores selling it in the US.
- I believe hot pepper sauce is primarily a male obsession and males are heavy users of the internet.

'What finally convinced me we could do it was the introduction of the 'Shop in a Box' concept. These are now widely available but I examined offerings from FreeServe MarketPlace and VirginBiz.

'The advantages of these packages are:

- They provide a hosting service, cart software and credit card processing all in one.
- Startup costs are low, typically less than £500.
- Running costs are low, typically less than £100 per month.
- Site design is done online using pre-built templates.

'The disadvantages are:

- The site design is limited – though to some extent that forces you to concentrate on content rather than gee-whiz effects.
- Some packages have minimum contract terms.
- Some of the other sites providing shopping links don't like linking to competitors sites. I was refused a link by one big name site because I am on a FreeServe Market Place.

'We chose to use FreeServe Market Place because the startup costs were low, there was no minimum contract period and we expect to get passing trade by being three clicks away from the homepage of a major portal like FreeServe. Now several problems had been resolved in one fell swoop and it was just a matter of eliminating the rest of the problems one by one.

'We registered Pepperama.co.uk with Nominet and use a redirection service to take people to FreeServe Market Place. That way we have our own identity and a place on FreeServe. We redirect Email to our own domain name in the same way. The cost was around £30 and very simple to organise. We were going to import sauces directly but found a UK distributor with a good range of produce – though importing is still an option. We found a local packaging company who delivers free and satisfies all our packaging needs. The Royal Mail offer a Daily Packet Post service which collects orders daily and delivers next day – they have been key to our business. Business banking and insurance were simple to arrange.

'The decision to go ahead was made in the Easter holidays; by 1st June we were in business.

'My wife runs the business, I do the technology and sauce tasting in my spare time!

'Business was very slow to start with – it's all too easy to assume you launch a website then the customers come from nowhere. It doesn't work like that.

'We have had to get low priced publicity wherever we can. This has included:

- Registering with as many search engines as possible – we've tried for over 300 so far.
- Registering with shopping sites.
- Finding magazines and papers with internet shopping columns and sending press releases and samples where requested.
- Attending trade shows
- Pepperama T-shirt
- Business cards given out at every opportunity
- FreeServe have recently run a Bonanza week promoting their position as a shopping portal.

'Business is building, slowly but surely. Most customers are men but there are an increasing number of women customers who are buying our products for their partners as presents. We are getting some very positive feedback.

'We are selling a product which fills a niche market, we provide the buyer with as much information as we can, there are no hidden extras, we deliver next working day and if there are problems, we resolve them quickly and efficiently.

'We aim to sell a premium product, with a premium service for a premium price – and have some fun along the way.'

20 July 2000

Analysis: the top and tail of a fruit B2B

Question and answer with one of the most dynamic American innovations in the sector.

Buyproduce.com [http://www.Buyproduce.com] is one of the leading B2B sites in the world. Here we give a detailed breakdown of the business and where it is going:

Q. How are transactions completed?

Suppliers build commodity lists and have the option to attach a number of price sheets to each list. The price sheets can be tailored to one specific buyer, many buyers, or all registered buyers on the system. In turn, buyers log on to the system and can see pricing for the products they are looking to purchase (there are no auctions involved nor will buyers ever send a Request For Quotes – RFQ). Buyers then submit a Purchase Order to a supplier for the products they wish to purchase. On the purchase order buyers have the opportunity to request price changes or add a detailed memo with regard to the shipment, i.e. 'please keep avocados in a gas room until shipping.' Suppliers can then accept or reject any orders. We offer customized features such as a transactional pricing model, which allows for streamlined communication between buyers and sellers.

Q. What other user services do you have?

Everything we do at Buyproduce.com is driven toward our ultimate goal of providing a total, multifunctional solution for produce buyers and sellers that enhances their business performance at all levels. For instance, the company is in the process of building a database with its customers, who will be able to access information for planning and category management purposes. This is just one of the ways we're expanding our capabilities to allow customers to use our solution in a multifunctional way, from invoicing to accounting. We're planning to incorporate additional site features such as logistics and financial services that will put us much closer to that goal.

Q. When did the idea of Buyproduce.com originate and how different is the present system from the original idea?

Following a family tradition steeped in agriculture, the early 1990s saw Buyproduce.com(tm) founders Brent Keefer and Bob Bonanno involved in the buying and selling of produce. Only in their 20s, the two typically worked all hours of the day calling and faxing potential buyers and sellers to complete orders. Acquainted on the phone when Keefer began selling Bonanno peppers and onions, they knew there had to be a better way to buy and sell produce. The two began to discuss their ideas of using the Internet to facilitate produce transactions and decided to meet. Within a year their concept becomes a reality as the Buyproduce.com (tm) transactional website is unveiled at the 1999 Produce Marketing Association's annual conference.

Q. What do you feel are buyproduce.com's main competitive advantages?

- Customer Service Focus
- Technological expertise
- Transactional Model allows flexibility to maintain and enhance current relationships and build new ones
- A company managed by produce people who understand the solutions for the produce industry, now and in the future
- Blue-chip funding for full-scale customer value initiatives
- Organisational depth
- Live since October 4, 1999
- On time and on target.

Q. You have been very aggressive in hiring people well known in the produce industry. How would the skills these people have brought to Buyproduce.com be ranked in importance as affecting the ability of other firms to enter the market?

Buyproduce.com was developed and is being managed by produce and retail veterans, who understand produce marketing and how it can be empowered by the Internet. The core distinction of the Buyproduce.com senior management is their domain expertise. As a whole they total over 120 years of produce experience ranging from the production side, as grower/shippers, to the retail side. Through their internal knowledge and hands-on experience they have the ability to foresee the potential effects that e-commerce can have on all levels of the produce industry. Being recognized as leaders in the industry has also helped to confirm to the industry that e-commerce is a solid and promising transition for the produce industry as a whole.

Q. Where does buyproduce.com see itself in the marketing channel; are you trying to take the place of the traditional terminal market? What do you perceive as your added value to the marketing channel?

Buyproduce.com links buyers and sellers through e-commerce, creating a business-to-business digital marketplace, focused on the unique requirements of the produce industry. Buyproduce.com serves as a supplier-buyer catalyst for providing one-stop, comprehensive fresh produce services, products, and information by utilizing resource aggregation, market research, and personalized service via an e-commerce vehicle. Buyproduce.com is an industry innovator, better established, and more focused on client needs than all known competitors.

Currently, the supplier side of the fresh produce industry is deeply fragmented, with key suppliers emerging only among specific crops. Buyproduce.com serves as a consolidator by gathering various suppliers together to conduct business in one web-based marketplace. This consolidation method is a proven concept with successes in other industries (i.e. books, financial services, etc.). Despite these proven examples, a lack of technology standards, engrained, archaic business practices, and rampant inefficiency have for years conspired to keep the fresh produce industry from modernizing and streamlining for its betterment. Now buyproduce.com has emerged with the vision, expertise, and execution required to lead the fresh produce industry into the new millennium.

Q. Will you be able to compete against growers who sell direct?

Most seller-owned websites make it difficult and time-consuming for a buyer to efficiently search for produce. We have developed the State of the Market(tm) feature, along with Compaq and Microsoft, as a highly advanced digital marketplace enabling buyers and sellers to post as many prices, update as often, and send to as few or as many customers as they want, all while being provided with a detailed analysis of an individual market. Furthermore, though basic websites have become very easy to create, developing a transactional site is extremely costly and complicated. The produce industry will experience more efficiency through 'one-stop' shopping and industry-wide standardization provided by buyproduce.com.

Q. What are the fees you're charging sellers?

Becoming a member is free. Sellers are charged a 1.5% marketing fee for each transaction. However, fees are always capped at $0.25 per carton equivalent. Buyers will never be charged a fee. As for additional revenue, our business model has been designed to extend into providing logistical services, credit/capital for transactions, as well as eventually offering a digital marketplace for other perishables, including meat, poultry, dairy, seafood, and floral. We also plan to offer several other value-added services including, but not limited to, data retrieval for customers through reporting, which will assist them in several areas including category management and tracking their own buying practices. There are also plans to begin selling advertising on our system.

Q. How many 'partnerships' on both the selling and buying sides, have you formed to date?

The company has been fully operational and actively facilitating produce sales since October 4, 1999. Currently there are more than 900 registered site users, including some of the largest suppliers and purchasers of produce.

Q. What is the list of produce you're seeing marketed at this time?

Our customers are currently trading thousands of various commodities on Buyproduce.com.

Q. What was Casio's role?

In an on-going effort to introduce new and emerging technologies, Buyproduce.com has developed an enterprise mobile data collection software that works on the Microsoft(tm) Windows CE operating system. The software was designed for the Casio E-105 using the Casio Compactflash camera. Its ability to process digital images in 65,000 colours is a vital part to our solution.

Q. Does your PRODUCEDIRECT software allow for any sort of data exchange with your users' existing software?

Through PRODUCEDIRECT(tm) technology, both buyers and suppliers can control their individual direct relationships. Members can maintain personal relationships through PRODUCEDIRECT's(tm) ability to allow one-to-one, one-to-many, and a one-to-all relationship. SUPPLIER – PRODUCEDIRECT(tm) allows suppliers immediate control over the information displayed about their products and their own market. A supplier can customise information for an individual buyer, a group of buyers, or for all buyers. PRODUCEDIRECT(tm) allows existing relationships to prosper with greater efficiency, while also allowing the supplier to enter into new relationships. BUYER – PRODUCEDIRECT(tm) lets the buyer determine the amount of information they view at a given time. PRODUCEDIRECT(tm) allows for searches to be as defined or as broad as a buyer may need.

Q. And on data exchange and EDI capability?

Buyproduce.com has announced the ability to facilitate the exchange of information between customers using EDI, ERP and other legacy systems and those using flat file communication. By implementing eVision, a TIE Commerce product (Formerly St. Paul Software), into the Buyproduce.com website, retailers and suppliers will be able to communicate with their existing technical resources, and achieve a higher degree of integration.

Buyproduce.com has created an interface that allows for direct transfer of information to Visual Produce – an accounting software package used by several companies and associations in the produce industry. This will allow Buyproduce.com to transmit information directly into a customer's internal accounting system – customers won't need to manually enter orders placed on buyproduce.com. Orders placed on buyproduce.com will automatically be transmitted to and updated on a customer's Visual Produce accounting system. Through this seamless integration, customers can continue to enjoy the benefits of Visual Produce while having the freedom to broaden their access to other trading partners using buyproduce.com.

Q. Has the handheld scanner unit specifically helped to convert buyers to your system?

The handheld device has not yet been made available to our customers. We have several units that we use for demonstration purposes, but have not begun offering it to our customers. However, the following are some of the benefits we have outlined:

- Assist shippers and receivers with order management and the documentation of discrepancies related to order management processes.
- Streamline the shipping process at the shipping location.
- Streamline the receiving process at the warehouse destination.
- Improve accuracy and efficiency in inventory applications. Information can be updated into internal inventory applications by attaching the hand-held to a dial-up cradle, or via a wireless connection.
- Allow receivers to record visual images of produce under consideration for rejection and, with the digital camera, can transmit images to all trading partners involved in the transaction for further evaluation and reconciliation.
- Enable growers to upload real-time images of the crop along with quality information from the field to potential buyers utilising the buyproduce.com digital marketplace.
- Give buyers, worldwide, access to up-to-the-minute information from digital images and make informed purchase decisions over the web.

Q. Are you releasing sales data only to members? Do you have any publicly reported summaries?

All data is the sole property of each customer, and will only be made available to the customer.

Q. To date, what has been the key to Buyproduce.com's growth and survival in this fledgling industry?

- First mover advantage – Customers of Buyproduce.com have faith in the honesty of our staff. buyproduce.com has always stayed true to its technical deadlines and has never made empty promises.
- Strong partnerships – Through our relationship with Microsoft and Compaq, buyproduce.com's system architecture was developed.
- Industry Leadership – Our senior management team is comprised of top produce industry leaders, who were the first to recognise the advantages of the digital marketplace for the produce industry.
- Integrity – All orders remain in the customer's private, secure system providing instant access to details only to that customer. Buyproduce.com will never talk about their customers unless they receive consent from that customer. Furthermore, all information is private.
- Innovation – The system provides customers with a clear, organized description of all ad requests, confirmations, and active ads. Using the State of the Market feature, buyers and sellers may post as many prices, update as often, and send to as few or as many customers as they want, all while being provided with a detailed analysis of an individual market.

Q. What are your main competitive concerns?

Our main competitive concern does not come from our existing competition; in fact, we currently do not recognize any of the current produce e-commerce sites as competitors. Our concern comes from the possibility of an outsider coming in and gaining domination of the landscape. This type of stealth competitor would be of a calibre equal to that of Oracle, VerticalNet, or CommerceOne.

20 July 2000

B2B portal brings in farmers

A goliath B2B site which only opened at the start of the year has formed an alliance between manufacturing and farming across Europe.

efoodmanager.com [http://www.efoodmanager.com], which is claiming to be Europe's leading business-to-business marketplace for the food industry, has announced an alliance with farmpartner.com [http://www.farmpartner.com], a sector-specific virtual marketplace geared towards agriculture.

The efoodmanager Open Marketplace is available to all subscribers to post their goods and services and also to request quotations and put out tenders. Meanwhile, efoodmanager is building Individual Company Marketplaces tailored to the individual needs of purchasers where the purchaser trades only with pre-selected third parties using his or her own product catalogues.

Daniel Kellmereit, efoodmanager's Head of Corporate Development, commented: 'The partnership with farmpartner.com opens up an exciting new sector for us and is a valuable partnership to add to our growing alliances. This cooperation allows companies within the food and agriculture industries to source producers and suppliers in a cost effective and time saving manner.'

Founded in January 2000, efoodmanager already has over 500 customers regularly using the company marketplaces, and a further 900 companies have registered to trade on the Open Marketplace. The company emerged from Fruchtnetz GmbH which was founded in 1997 and was the first European Internet marketplace in the food industry. efoodmanager currently has 50 employees from the food, e-commerce and consultancy sectors. It has offices in Duesseldorf, London, Barcelona, Paris, New Jersey and New York. Farmpartner.com was founded in February 2000 and currently employs a team of 25.

Efoodmanager's initial service will focus on the fruit & vegetable and seafood industries. Marketplaces for meat, meat convenience products, poultry, deli products and dairy products will be launched in the coming months. Initially accessible in English and German, each marketplace will contain over 25,000 product categories. Efoodmanager plans to launch further marketplaces for beverages and food commodities such as grains and ingredients in the near future.

Jan Stenger, co-founder and Managing Director of efoodmanager, commented: 'Our new online environment has created a marketplace for buyers and sellers that has not existed before. Not only can the international food industry be accessed cost effectively by European buyers, but process and product costs can be reduced by up to 25%. Our goal is to totally integrate value added services with the trading processes so as to deliver the best value for our customers.'

Buyer advantages of the electronic purchase of food produce over the efoodmanager platform include: quick and simple information retrieval, a rapid transaction process, access to a large pool of suppliers, price reductions resulting from greater price transparency, individual statistic tools such as sales reports, average prices and trend analyses – and a soon to be launched tracing and tracking service.

Advantages for selling food products over the efoodmanager marketplace include: access to a large buyer market with dramatically reduced customer acquisition costs, payment security, use of supplementary services, cost reduction for marketing and sales activities, speed of 'time to market' and the ability to respond rapidly when there is volatile demand.

Targeting the procurement and trading needs of the wholesaler/trader, retailer, producers and food-service companies, efoodmanager is initially free to join. From the fourth quarter of 2000, efoodmanager will begin generating revenues by charging small transaction fees for every trade processed through the efoodmanager marketplaces. Efoodmanager will also take service fees for building and hosting the individual Company Marketplaces.

11 August 2000

Tradingproduce.com unveils field to Web technology

One of the biggest B2B sites in America has revealed its new software breakthrough for trading through the Web.

Field inspectors and logistics will now be able to up and download reports, digital images, bar code information and bills of lading in real time on the Net.

'People doing field inspections can download the format of reports they need to file, fill out the information on the report, and attach any images they need to,' explained TradingProduce.com [http://www.TradingProduce.com] Chief Technical Officer Zia Zahiri. 'The buyer can than click on the product in which they are interested and immediately get the report with real-time images.

'At the shipping or receiving dock, the passing of the product can be confirmed, including the amount shipped and received and pictures of the goods, both before shipping and at delivery. We are enabling technology to make the flow of information faster, easier and mobile. This is another step to eliminate the paperwork involved.'

Mobilbridge software allows for the capture of bar codes and digital images, video clips, and sound bytes as well as text. Its patent-pending funnel technology automatically associates diverse input streams, such as images from a digital camera, with the appropriate fields on a web page so the information can be uploaded. When the connection is made to the TradingProduce.com portal, all data submitted offline – text, images, video and audio – can be transferred by merely clicking the submit button on the hand-held device, or PDA.

'Inspectors in, for example, the Salinas Valley can do field reports that can be viewed in real time with digital images of the product in Europe,' said TradingProduce.com Chief Executive Officer Rob Bonavito. 'This value-added capability is available now to allow our members to have instantaneous access to field inspection reports, loading information and transportation.'

TradingProduce.com's service- and solution-oriented website charges its members no buy/sell transaction fees and provides an opportunity for significant savings throughout all the perishable food industries.

It also offers the opportunity for the customer to move the entire transaction process electronically through the Internet to their own accounting system with no paperwork involved. Its business model is designed to create revenue from value-added services, including logistics, financial services, analysis tools and engineering.

It is backed by the $950 million investment fund Bay Isle in San Francisco, Silicon Valley private investors and vTraction, an affiliate of Rabobank, which is one of the world's largest and strongest financial institutions with $290 billion in assets.

01 September 2000

Analysis: Cooking on line

2500 sign up for cookery courses.

Recipes were, along with sex, one of the first subjects to populate the Net and have become one of the more sophisticated areas of web surfing. In America, cooking online is one of the biggest growth areas of e-commerce. This summer, Cook's Illustrated magazine's Cook's University launched classes such as Everyday Sauté. For $40 subscribers get to learn the intricate balance of flame, butter, oil and pan. Classes last from two to four weeks, and the registration fee includes recipe books and the opportunity to question instructors. Because classes are accessible for 30 days, students can pace themselves, working according to their own schedules. E-mailed questions are answered generally within 24 hours, and students are encouraged to discuss the material electronically among themselves.

About 2,500 people had taken the online classes at Cooksillustrated.com [http://www.Cooksillustrated.com] as of late July.

While there are other opportunities on the Internet for cooking instruction, most do not provide for interactive communication.

The New York Times' website Nyt.com [http://Nyt.com] offers streaming of video images and instruction on cooking specific dishes in cooperation with the French Culinary Institute. For example, you can learn how to cook a whole fish by watching it being done, and by listening to a chef describe the technique. There also are recipes and text instructions available by clicking on handy links.

Tavolo.com [http://www.Tavolo.com], formerly known as digitalchef.com, offers recipes for complete meals from appetiser to dessert. The site also lets you search its database for recipes containing the ingredients you have at hand. Got tomatoes, garlic and fish, nuts and fruit? Click on the boxes next to these ingredient choices and see what comes up.

Cookware maker Corningware invites consumers to 'find your inner chef' at its Corningware eSchool at Corningware-eschool.com [http://www.Corningware-eschool.com].

Corningware's parent company World Kitchen Inc. developed the venture with the College of Culinary Arts at Johnson & Wales University of Providence, Rhode Island. The eSchool offers step-by-step instruction with photographs, and covers a variety of techniques, including baking, grilling, broiling, braising, roasting, stewing, poaching and steaming.

'We started it in November as part of a whole programme we were doing on the Corningware brand to target younger consumers,' explained Lynne Recktenwald, World Kitchen's vice-president of oven and bake ware. 'We wanted to teach the fundamentals – the things you take for granted that everybody knows, but they don't, really.'

Corningware officials boast they're 'wiping out cooking illiteracy,' Recktenwald said.

At Loadtv.com [http://www.Loadtv.com], online cooks can watch video of food being prepared, with accompanying instruction. Special software must be downloaded first, however.

Foodvision.com [http://www.Foodvision.com] visitors can click on a link and watch any of 18 cooking shows. The 20-minute slide shows feature such recipes as pork medallions pocketed with apple-cherry stuffing and served with mashed pecan sweet potatoes.

'I would like to do smaller segments now,' said Foodvision's Corporate Executive Chef Geir Kilen. 'Knife use, sharpening, dicing an onion – maybe two- or three-minute segments.' Such bite-sized videos are just a future plan for now, but Foodvision's longer shows remain 'viewable at any time,' Kilen said.

Most if not all of the sites have a marketing component, offering cookware, cookbooks and other opportunities to buy merchandise. And Cook's University hopes to angle more subscribers for Cook's Illustrated.

05 September 2000

Interview: USA model flourishes quietly in the UK

A small gourmet site dedicated to specialist companies has steered a shrewd course around the storms of Net trading.

Borrowing a name from one of the biggest portals in the world and a template based on an American beer site realbeer.com [http://www.realbeer.com], the small and enterprising Aboutfood.co.uk [http://www.Aboutfood.co.uk] has kept itself afloat from web authoring revenues and let the advertising start to build up. Nick Godwin and Nick Henley started the site at the end of 1999 'as a response to the rather disparate nature of information on the Internet relating to good food in Britain – in particular where to find and buy speciality food and drink outside of London'.

Godwin explains: 'An acquaintance, Pat Hagerman of American beer portal realbeer.com, provided the successful publishing system they had developed over four years and the last three or four months of 1999 were spent putting together a database of over 4000 suppliers, from farm shops and organic delivery schemes to delicatessens and online retailers. A plethora of different guides, brochures and fliers were consulted and companies were subsequently mailed in order to clarify details.

'Inspired by the realbeer model, the plan was to create an information resource site focused on fine food and drink (particularly the food side) and I came in at the beginning of 2000 to manage the content side of things. The cornerstone of content on our site is our library. This is an archive of food writing focused on fine food in Britain, allowing people to search for articles about anything from jam making to restaurants in Surrey. We felt that many good articles appear in magazines or newspapers for instance which once published are often 'lost' to all intents and purposes. It is also a showcase for writers, both established and emerging, as well as niche and regional publications with good food writing content.

'We launched the site in late April to a flurry of technical problems – the task of adapting a publishing system tailor-made for someone else's website to what is a very large site caused more problems than had been anticipated. It wasn't for another couple of months before everything was working technically, and we had started to evolve our own look and feel. Obviously this was a strain on the site financially, but this spawned Orchard Internet, our sister web design and marketing consultancy company catering for larger corporate clients (online marketing consultancy which had been Nick Henley's background) and the growth and success of this venture has helped support About Food's growth without the requirement for large external investment, a fact that, in the light of the crash of other recent ventures, has left us in a pretty healthy position.

'I am now responsible for all of the day-to-day running of the site with part-time assistance from the seven Orchard Internet staff. Its been a hectic last five months which has seen constant growth and improvement of the site and our traffic rise greatly to the point where we are now an attractive proposition for advertisers. We are aiming to grow organically, building at an achievable rate.'

17 November 2000

Analysis: the truth about Internet fraud

The American Federal Trade Commission has finally released figures showing where the key areas of Internet fraud are taking place.

In a top ten of frauds, the commission has revealed the most common complaints based on a total of 285,000 complaints. It claims to have successfully prosecuted 251 online scammers. 'We want the 'dot-con' artists to know that we're building a consumer protection coalition that spans the globe. We aim to make the Net safe for consumers,' said Jodie Bernstein, Director of the FTC's Bureau of Consumer Protection. 'This collaboration with law enforcement agencies, industry and consumers will create a climate where e-commerce can be conducted with confidence.' The biggest scams are:

1. Auctions

The popularity of online auctions has led to explosive growth in fraud, with the number of reported cases ballooning from 100 in 1997 to a whopping 10,000 in 1999. The average victim loses $293 according to the Internet Fraud Watch, usually paying for goods that are never delivered.

2. Fraudulent billing

Three of the top ten Web scams involve fraudulent billing. For example, in the No. 2 scam, ISP fraud, a company will fraudulently sign up consumers for Net access and then put phony charges on their monthly phone bills. In one case, scammers mailed 'rebate' checks for $3.50 to consumers. When people cashed the checks, they unwittingly 'agreed' to allow the defendants to provide Internet access services. The defendants then started charging the consumers between $19.95 and $29.95 a month and 'made it nearly impossible to cancel future monthly charges and receive refunds.'

3. Web cramming

In a related scam, scam artists call small businesses and offer 'free' Web pages. After the victim signs up, the scam artists begin charging high fees.

4. Net porn credit card fraud

The No. 4 'dot-con,' Net porn credit card cramming, targets the Web surfer who gives an adult website his or her credit card number because the site says it needs the number to verify that the visitor is of age. After obtaining the credit card data, the scam artists start billing the credit card without authorization. The FTC recently sued the operators of three adult-oriented websites allegedly engaged in this practice and asked the court to freeze the assets in two of the cases.

5. Investment and 'get rich' schemes

Preying on the hopes and dreams of individuals who want to make a little extra cash, some of the less common Web scams promise victims 'the moon and the stars,' but deliver only heartache.

6. Pyramid selling

In the 'pyramid scam,' the con artists promise people they will make money through selling products or services, as well as through the products sold by people they recruit into the program.

7. Invest to get rich

Some Web con-artists, realizing that not everyone will fall for the promise of 'something for nothing,' resort to offering people the chance to invest in business opportunities or home-based moneymaking programs.

8. Travel scams

Companies offer consumers luxurious trips with 'extras' at bargain-basement prices. Some of these web scam artists deliver lower quality accommodations and services than promised, while others fail to deliver any travel services at all.

9. Web versions of real-world scams

A variation on the telephone billing scam. In many of these cases, the con artists offer consumers free access to adult material if the consumers will download a 'viewer' or 'dialer' computer program. What the scam artists do not tell consumers is that once they have downloaded the dialer, it will disconnect their modem from the Internet and then reconnect through an international long-distance number, resulting in exorbitant long-distance charges.

10. Miracle cures

The last scam on the FBI's list of the 'Top Ten Dot-Cons' is a modern day version of the snake oil salesman. In these scams, e-tailers sell consumers 'miracle' cures that are 'proven' to cure serious and even fatal health problems.

17 November 2000

Part 4
E-Commerce

Business2Business has already begun to change the way we work. If email was the first step, then the really big changes are just around the corner as the Internet shapes up to take over traditional ways of working.

He is 30, married and well educated . . . but she is catching him up

The typical Internet shopper is male, well educated and makes more than £30,000 a year; but look out: his wife is catching him up, says a new survey.

The typical British Internet shopper, according to a new report from Ernst and Young, is younger, with two thirds of those surveyed under 40. Over half had household incomes of around £30,000 and education levels were high.

Two thirds of those surveyed were men, half of whom were married, but Ted Yates, head of consumer and retail, said the number of women e-shoppers was increasing. 'Men are typically first to embrace electronics, but women are beginning to take their place driving the mouse – as opposed to the car – as they already have done in the US,' he said, 'Just as they dominate shopping on the high street so they are beginning to assume a stronger presence in the virtual world'.

10 January 2000

UK Net usage surges ahead of France and Germany

One in five UK homes accessed the Internet to shop in December compared with less than 10% in Germany and France.

Nearly 21.2% of the U.K. population, or 10.1 million people, accessed the Internet from home according to new research from a survey of 2985 by Taylor Nelson Sofres in December 1999.

Men represented the biggest percentage of users at 58.5%. Women represented 41.5%, up from 37% in November. Users in the 35-49 age group accessed the Internet the most, with 32.3%, followed by people under the age of 25, with 24.8%, and people aged 25-34, with 22.5%. The over-50s accessed the least, with 20.4%.

The proportion of UK home users that visited an online shopping site rose to 65.8% from 61.7% in November, as the Christmas holiday season led to an increase in retail buying. Books, compact discs and movies were the most popular items bought online.

The website most accessed by UK home users in December was predictably Microsoft Network, followed by Yahoo!, America Online, and Freeserve.

Users across Europe most often access the Internet to surf the Web, followed by sending and receiving e-mail, downloading data and software, and viewing or accessing video and audio files.

17 February 2000

Language slows up European growth

New research puts the UK well ahead in the e-commerce race while other European countries are still mistrustful of buying online.

Non-English speakers cannot always find sufficient websites in their language which offer familiar products, Forrester Research analyst Reineke Reitsma said. In France, companies trying to promote online sales have an additional obstacle in the form of competition from Minitel, the country's widely used teletext service, she continued.

The survey in July 1999 and August 1999 covered 17,000 households in Europe's three largest markets (Germany, Britain and France) as well as Sweden and the Netherlands. Christmas figures have since updated this with some 18% of the UK population now having bought online but with no such increase in continental Europe.

It found that only 18% of European homes had a connection to the Web at home and less than 10% were actually prepared to buy anything online. In Sweden, where e-commerce was most popular, 14% of households had visited company or product websites, and 7% ordered goods. At the other end of the scale, 7%, or around 1.8 million, of France's 26 million homes were linked to the Net, and only 2% had bought anything.

By contrast, one-third of French homes, or 8.6 million, subscribe to Minitel, which has offered them shopping and information services for 17 years. The service generates a turnover of 5.46 billion francs.

As well as being reluctant to put personal information on the Internet, two-thirds of European shoppers preferred to see goods in the shops, the Forrester survey found. That means companies wanting to sell goods over the Internet need precisely targeted marketing strategies and must do more to encourage surfers to shop, Reitsma said. E-tailers need to reassure potential Web shoppers that their personal and financial details are safe and ensure their goods will be delivered efficiently, she added.

As it can take two years before a Net surfer becomes a Net shopper, e-tailers also needed tactics to keep potential buyers coming back to their site. For example, Swedish furniture chain IKEA has a website where visitors can change the lighting in a virtual house to see what effect they preferred before committing to buying. A Dutch paint firm lets its website users experiment with different wall colours.

While low prices were not the main reason European consumers gave for purchasing online, convenience was cited as the main attraction, it is becoming an issue in Britain and Sweden and will become increasingly important elsewhere. Webshoppers in Europe were most likely to be people wanting to improve their careers, followed by those seeking entertainment or goods for their families.

22 February 2000

Online farming site goes for £7 million

One of the longer standing farming sites in the UK has been given £7 million of investment to move into Europe.

The Internet Capital Group Europe said it did not believe the supply chain to farmers was as efficient as many people thought. FOL is a traditional portal play combining news, community resources and e-commerce. 'We started back in 1996 as an information provider, then set up an ISP that targeted farmers', said marketing manager Jon Swaby. Farmingonline.com [http://www.farmingonline.co.uk] offers a hosted transaction service to around 100 farming-related companies, based on a homegrown e-commerce platform. The site targeted the top 20% of the farm business market in the UK, Swaby said. It claims 1.5 million impressions a month and a registered user base of 18,000.

'We want them to expand very rapidly into Europe, initially Germany,' said Stephen Duckett, managing director of Internet Capital Group Europe. Duckett said Internet Capital Group had made several investments in the more developed US agricultural portal market, including Emergeinteractive.com [http://www.emergeinteractive.com], a cattle trading site that recently went public on the NASDAQ, and Cybercrop.com [http://www.cybercrop.com]. Competition is likely to come from Reed Elsevier's Farmers Weekly publication and also the Atlas Venture-backed Agrifirst.com [http://www.agrifirst.com/], a French start-up.

12 March 2000

Brake Brothers to invest £12 million

The biggest catering supply company in the UK, Brake Brothers, is to invest £12 million to create an online ordering system for its customers.

The web-based ordering system will be available for all Brake Brothers plc [http://www.brake.co.uk/]'s 100,000 customers, bringing e-commerce directly to pubs across the country. The move is expected to slash jobs in the 500-strong telesales department.

16 March 2000

Macdonald's buys into Food.com

Macdonald's and Kraft Foods are two of a consortium of American conglomerates putting $80 million into the site.

Food.com [http://www.food.com] started as a home delivery service for restaurants in the Texas area and has recently branched out into online bookings to top-class restaurants in cities like Boston. This will be Macdonald's first move into e-commerce. Other parties will include non-food groups such as Blockbuster, and Liberty Digital – a division of Liberty Media.

A spokesman rather vaguely said the plan was to 'create a sort of portal to all things to do with dining out'.

16 March 2000

New B2B food portal gets £2.6 million venture-backing

The lure of the £500 million European business-to-business market will see the launch in May of a new portal site.

eFoodmanager.com [http://www.efoodmanager.com] will connect thousands of food suppliers with bulk buyers such as supermarkets, armies and hotel chains. It will be in five languages and with 85 partners including Tengelman, the German retail group.

Jan Stenger, joint managing director, said that eFoodmanagers hopes to have at least ten bulk buyers and 500 traders on the platform by the end of year. Within three years it wants to capture 10% of the £500 million European market for food trade.

He said the service would cut processing and production costs by up to 30%. Alexander Bruehl, principal at Atlas Venture who will join the board of eFoodmanager, said: 'The food industry is one of several market sectors where the ineffective, slow processes of current trading are about to be revolutionised by the Internet.'

16 March 2000

Grocery market to hit £2.3 billion in next four years

Online sales of food which have been languishing in the relegation zone of e-commerce are set to leap to the front of the e-commerce revolution, according to new research.

A study from the UK-based Verdict Research group, says the online world is to follow in the footsteps of the real world and have groceries as the bread-and-butter provider of funds for e-commerce. By 2004, Verdict predicts, online groceries will be the most commonly bought item on the Internet. The study estimates that it will be a marketplace worth around £2.3 billion in the UK alone, compared to the current £165 million. And that by this time 3% of all sales will take place on the Web.

28 March 2000

Nestlé and Danone tie up for e-buying

The much-predicted arrival of e-procurement across Europe has begun with the two giant food companies joining forces to create a formidable buying network directly over the Net.

Groupe Danone, one of the world's biggest suppliers of food products has linked with Swiss giant Nestlé to set up a European Internet exchange for procurement of consumer goods. The new company will be called CPGMarket and will be launched in July. It is in competition with the exchange launched by a group of 50 US-based consumer products companies, including Proctor & Gamble, Kraft Foods, Bestfoods and Colgate Palmolive.

28 March 2000

500,000 credit card details go AWOL

After the revelation of the biggest credit card fraud in Net history, Barclays bank announces a new secure payment programme.

In the largest known case of cybertheft to date, a computer intruder stole information on nearly half a million credit cards from an E-commerce site, and then stored them all on the database of a US government agency.

Cards involved were from Visa, MasterCard, American Express and Discover. The biggest known case previously was of 300,000 details from the CD Universe site. The latest case emerged from a two-day national summit in the US called by the Treasury department to identify theft and to focus on the growing threat.

Barclays is launching a secure purchase-to-payment e-procurement system which it says will provide a direct channel for the sale and delivery of business and financial services. Barclays B2B.com will bring in Andersen Consulting to help design the service, and Oracle to provide procurement software.

E-business and software service group Parity is launching a service aimed at companies who want to send UK customers personalised billing on the Internet.

28 March 2000

Europeans like online shopping

While American consumers are still walking the aisles of their supermarkets comparable figures for Europe show an impending bonanza for e-grocers.

By 2010, grocery shopping in Europe is expected to reach about $100 billion, or 10% of all grocery shopping, according to a study conducted by a European management consulting company, Roland Berger & Partners Inc.

The study, which was based on interviews with 6,000 customers in six countries, found that Europeans like buying online. Reasons for the preference include the fact that produce is fresher and that it is handled less frequently: the average supermarket apple, for example, is touched by 11 people before purchase, according to the study. The study also found that time pressures and rising stress levels, particularly among working women, are also boosting online grocery buying.

07 April 2000

Internet Trading Exchange for food industry

Two leading food distributors are linking up to form a business-to-business online exchange for non-consumer goods and services.

Food Distributors International and the Food Marketing Institute have agreed the deal that will link foodservice members representing the $800 billion North American food industry. The two associations are planning to launch the site later this summer.

IBM will be managing the activities of the alliance partners, which will include i2 Technologies, Inc and Ariba, Inc. Christian Nivoix, general manager of IBM's distribution sector, said: 'E-markets are exploding. This alliance between FMI and FDI gives its members the opportunity to use Internet technologies to create new cost structures, to open new distribution channels, to transform competitiveness and change the industry. That's the real impact of the Internet.

'We see this as a dynamic electronic marketplace for our industry. It will enable us to deliver at Internet speed the B2B e-commerce, education and communications services that our members need.'

FMI President and CEO, Tim Hammonds said: 'We believe this set of technology partners will enable us to deliver state of the art Internet buying fully integrated with logistics services, coordinated backhaul, and cooperative planning, forecasting and replenishment.

'In other words, the Internet buying function is only the beginning. The real pay-off comes from integrating the total supply chain.'

14 April 2000

Euro e-commerce sales to top £105 billion by 2005

American research company Forrester has projected the size of the European market for e-commerce to break the £100 billion barrier within five years.

Meanwhile other research suggests that UK consumers will spend £10 billion this year alone.

25 April 2000

Europeans buck the trend

Another survey has underlined the growing feeling that the European grocery market may not follow American trend.

This survey from Roland Berger, a management consultancy, has shown that Europeans seem happy to forsake the weekly supermarket shop for the added freshness and convenience of Internet shopping. It also suggests that more Net grocers are showing a profit than had previously been thought and anticipates 10% of all grocery shopping moving to the Web over the next decade.

The survey results coincide with Tesco's announcement that it will create 400 new jobs by extending its home shopping through Scotland. And ironically the survey warns of a potential traffic problem in inner cities with the arrival of a new generation of home-delivery vans. Reading between the lines, the heartland for e-commerce seems increasingly established in dense urban areas, such as San Francisco and Manhattan, and that pattern seems to be recurring in Europe.

E-shopping is the most interactive of web pursuits and the weekly grocery shop seems to have established itself in the UK as an obvious conversion point for new surfers. This trend is supported by emerging statistics from European countries such as Germany where use of the Net is widespread.

Other countries like Italy and Spain which lag behind in PC usage are expected to access Net information via their mobile phones in the future, which may create a new barrier to e-commerce grocery shopping: for all the excitement surrounding Wap technology, few phones currently offer anything except rudimentary web access. The prospect of ordering a whole list of groceries on the key pad of a phone would be enough to put off all but the most ardent enthusiasts.

02 May 2000

Food portal swallows up fish industry site

A major new European food portal has underlined the seriousness of its intent with the capture of one of the biggest B2B sites in the fish industry.

eFoodmanager.com [http://www.efoodmanager.com] has acquired the German site DerFischmann.de, for an undisclosed figure. The founder and CEO of

DerFischmann, Joachim Reichelmann, also joins eFoodmanager as category head for seafood produce.

Founded in 1996 DerFischmann.de has some 100 seafood industry participants currently trading through the platform. Jan Stenger, co-founder and CEO of eFoodmanager, commented: 'The acquisition provides an important building block to access Europe's leading buying and selling groups. We plan to have all major food categories tradable on the platform by the year end.'

With 250 participants now signed to eFoodmanager's platform, including retail and e-tailer groups, manufacturer and traders, the company plans to pilot its service across four initial food categories in May:

- fruit and vegetables
- seafood
- meat and
- dairy products.

eFoodmanager.com will be divided into three main areas:

- the *Open Market* where buyers and sellers can trade openly or anonymously amongst themselves;
- *Individual Channels* where buyers trade only with preselected parties and
- the *Services area* where buyers and sellers can take advantage of the logistics, packaging, insurance, financial services, information, business directories and food produce directories managed by eFoodmanager.

As part of the Open Market there will be an auction area where members can sell off unsold or highly demanded produce or products.

eFoodmanager expects to capture 10% of the European food industry's online trading activity by 2003. Initially free to join, eFoodmanager will generate its revenues through transaction fees on products and services sourced via the site, as well as service fees for building and hosting the individual channels.

eFoodmanager currently employs 30 people from across the food, e-commerce and management consultancy industries, and has offices in Austin (Texas), London, Dusseldorf, and Valencia.

02 May 2000

New B2B site for organic

An ambitious new portal is to be unveiled this summer focussing on organic trading across Europe.

The site is being championed by Fergal O'Mullane of Finest Organics and will be called OrganicTrader.com [http://www.organictrader.com]. O'Mullane's background is in the organic meat business.

The site plans to audit suppliers and manufacturers by means of on-site inspections and then link them up into a European wide organic market place. Development work is by Citria who developed WorldofFruit.com [http://www.worldoffruit.com].

08 May 2000

It is not all rosy in the B2B area

Further gloomy predictions of market shakeouts have extended to the B2B arena.

A new report suggests many B2B trading exchanges will have to rely on advertising and give up their independence to survive.

Boston, Massachusetts-based AMR Research predicts that the number of exchanges will shrink from today's 600 to fewer than 100 firms by 2001: 'Independent trading exchanges are finding it difficult to drive business,' said AMR analyst Scott Latham in the report. 'Most have not processed their first transaction.'

In fact, the report points out that even some of the top-ranked exchanges are finding that most of their revenue comes from advertising. AMR is predicting that only two or three B2B ventures will survive in each industry to claim their share of the e-commerce pie.

Some sites are choosing to consolidate with competitors, while others are following Amazon.com's lead by expanding into new markets with additional product offerings. Some exchanges are choosing to give up their neutrality by partnering with existing brick-and-mortar companies to gain funds, validity, and transaction volume. Others have given away equity in return for promises from buying organisations to sole-source through them.

The problem is, according to Latham, that the e-marketplaces are being driven to accept deals that offer heavily discounted terms or no revenue at all. Some exchanges are 'giving away the farm to anchor customers,' Latham said.

Suppliers are balking at the high transaction fees that some exchanges are charging, and are choosing to establish their own marketplaces on their own terms. These supplier-driven exchanges carry both the suppliers' own goods and related products.

08 May 2000

Welcome to the era of intelligent software

Before the year is out surfers face the advent of intelligent browsers that sort the Net out for them.

From all directions new innovations are coming online that will allow computers to get to know their masters and scoot off to bring back relevant information.

The latest breakthrough was announced by Experian, a data agency which is claiming a revolutionary new software that allows websites to recognise potential customers. This will then filter back to retailers and manufacturers and allow companies to target-market individuals. The biggest advantages will come in the area of fraud, by identifying genuine customers, and in niche marketing, by allowing much greater accuracy.

In the wings too is autonomy.com [http://www.autonomy.com/] which allows users to have their own private search engine customised to their preferences. Tesco announced last week that it was investing £11 million in technology including Autonomy's *My Web* software which will allow shoppers to make instant price comparisons with rival or related brands.

And from the mobile phone sector will come WAP messaging which can alert customers to deals within 200 yards of wherever they happen to be.

Not to be outdone Swatch watches is pioneering the watch as another means of connecting to the Net. Set to retail at £52 and £67 watches will be e-mail enabled and also allow Net surfing and messaging of tickets . . . Already major Swiss ski resorts have done away with old-fashioned lift passes in favour of electronic passes, which can also be programmed into the watch.

And later this year it is expected that the talking watch that works as a phone will be unveiled.

15 May 2000

It's business business business

E-commerce has now taken over as the driving force on the Net, according to the UK based Fletcher Research.

Fletcher finds e-commerce seemingly going from strength to strength in Britain. Rebecca Ulph, senior analyst, said UK corporations have put up strong competition to Web-based start-ups. She quoted Smile, the Internet bank started by the Co-operative Bank, as a prime example: 'one of the strongest brands in UK online financial services.'

Another example of the corporate giants arriving late in the market place but with huge financial muscle is the tie in of rival hotel groups Marriott and Hyatt in a separate company to connect buyers and sellers in the hospitality industry. This is touted as a new generation B2B site and will go head to head with Zoho.com which received $50 million in a third round of financing from leading technology groups including Dell and Ariba.

15 May 2000

Underlying trade in e-commerce is booming

Never mind market corrections, assets top $1 trillion for the first time.

The stock markets may be in a state of global flux, but according to the latest statistics the online trading world is rolling along faster than a Formula 1 racing car. According to US Bancorp Piper Jaffray, stock trades online surged a record

69% in the first quarter, and assets held in online brokerage accounts have surpassed $1 trillion for the first time.

15 May 2000

Advertising set to kick in as revenue model

Media marketer predicts advertising revenues are about to kick in.

William Wemyss of Panlogic told delegates to the Efood conference in Holborn London that Net marketing was likely to prove the most effective way of reaching targeted individuals. And he predicted that secondary sites could well soon benefit from a much needed cash injection of advertising revenues.

'With so many companies now receiving money from venture capital, it will be much harder for sites to have unconventional arrangements like transactional deals. Venture capitalists will demand conventional advertising. The fall out from Boo.com and Netimperative will if anything make the business model more critical. Venture capital will be paying much closer attention to the business model and not just the number of eyeballs.'

But he warned that the quality of most e-lists was still largely bogus. 'There is a huge potential to spam. And currently no one knows the value of any lists. So there is potential to damage your own brand down the line. Many of the lists on the market today are currently derived from North America with only perhaps 5% of the names being European, let alone British.

Wemyss advocated a programme of list building within the site using such vehicles as competitions to attract email capture.

30 May 2000

Produce B2B picks up $20 million new investment

First mover status has propelled a fruit and produce buying B2B site to the leading edge of e-commerce.

BuyProduce.com [http://www.buyproduce.com/] now has nearly 1000 registered users since launching last October. It has just received a new $20 million in injection of funding.

The site works on three levels: transactions, information and collaboration for collective marketing. The main focus though has been an unexpected commitment to adding value executive services such as back-end integration, electronic funds transfer, market research, credit services, and logistics. It has worked closely with Casio computers to develop handheld cameras that can film produce in the field real time to show to prospective buyers.

The new round of funding is led by Charter Growth Capital and includes investments from Lehman Brothers and Nassau Capital. The world market for fresh produce is thought to be around $200 billion. 'We are produce people working on behalf of produce people,' said Greg Flood, CEO of BuyProduce.com. 'At a time when venture capitalists are increasingly cautious about where to invest their money, this investment and endorsement is a testament to the BuyProduce.com business model.'

08 June 2000

World of Fruit claims 300 companies have signed up

A produce B2B site that was set up last September is now claiming 300 users across 25 countries.

The global produce market is currently estimated at $250bn. Analysts suggest as much as 20% of this will move on-line over the next three years. worldoffruit.com [http://www.worldoffruit.com] launched in September 1999, now has 300 registered users in 25 countries. Its internet market place lets buyers of anything from oranges to pears compare prices from produce suppliers before placing orders. The site deals in euros and is pitched at major purchasers. Sellers pay the site a commission on their sales.

With offices already in the UK, France, Spain, Italy and the Netherlands, World of Fruit plans to expand into the Americas, South Africa and Asia promoting the online market and offering customer support.

Business development director, Francis Hackett says: 'The internet is providing transparency and eliminating geographical boundaries. You put produce up on the site and it eliminates the need to phone or use the fax'.

03 July 2000

Sainsbury to put $15 billion of global purchasing online

Supermarket backs retail B2B site.

GlobalNetXchange [http://www.GlobalNetXchange.com], the major B2B online market-place for the retail industry, plans to give supermarkets access to suppliers of everything from food to condoms and chocolate. This $200bn internet venture is backed by Oracle and heavyweight chains around the world including Sainsbury's [http://www.sainsbury.co.uk].

'We can see that the whole way people buy goods within the retail market is going to change' said Sainsbury's group director for e-commerce, Patrick McHugh. Underlining the supermarket's eagerness to embrace the Net, he points out that it will be putting 75% of its global purchasing through the exchange which amounts to about £15bn.

McHugh compares the GlobalNetXchange site to a giant yellow pages where buyers can congregate and browse suppliers by categories. To preserve confidentiality, there are closed areas for striking the deals. 'When we find suppliers through the internet we go into normal commercial negotiations about price, delivery and quality through secure links.'

McHugh says Sainsbury's will still carry out offline checks on suppliers. 'We send buyers around the world to look at bananas and pineapples. There is no way through that but we will be able to do it in a more informed way.'

McHugh says the chain is strongly committed to finding new suppliers online. 'I believe the Internet will give us much, much more choice. It will make us more competitive in our own-label products and will enable us to broaden our range.' He believes that B2B sites are not just convenient tools for buyers but can really assist suppliers too. Letting them reach a number of major buyers in one location so they can consolidate their orders.

03 July 2000

Webvan buys HomeGrocer for $1.2billion

The American online grocery market has consolidated further with two of the biggest players merging.

The move is yet another step in Webvan [http://webvan.com]'s strategy to expand its geographic reach, along with moving beyond groceries to become an all-purpose Internet retailer. Earlier this month, the company added consumer electronics and videos to a product line-up that includes pet supplies, kitchen and bath accessories, books, home hardware, office supplies, hosiery and a 'smoke shop.'

HomeGrocer [http://www.HomeGrocer.com] went public in March at $12 per share, backed by blue chip and Internet names like Amazon.com, Kleiner Perkins Caufield & Byers, and the Barksdale Group. Revenues for the year ended January 1 rose almost 1,900 percent but HomeGrocer's net loss widened to $84 million from $8 million. Shares of both companies were lower after the announcement.

'Clearly, the online marketplace is evolving,' said Webvan President George T. Shaheen. 'By combining, the companies will reduce costs and improve finances. It is also a good fit from the standpoint of market reach, expansion, and fulfilment and distribution infrastructure.'

Webvan is projecting revenues of $300 million this year and an astonishing $1.2 billion next year.

Webvan plans to serve 13 markets by the end of the year: Atlanta, Georgia; Baltimore, Maryland; Bergen County, New Jersey; Chicago, Illinois; Dallas, Texas; Los Angeles; Orange County, California; Portland, Oregon; Sacramento, California; San Diego, California; San Francisco, California; Seattle, Washington; and Washington, D.C.

03 July 2000

Delivery charges holding up e-commerce

Pilot projects in Boston have revealed that customers do not like to pay delivery charges.

Speaking at the Etail 2000 [http://www.wbresearch.com/etail00usa/] conference, David Martin of Asda@Home, revealed that some American operators had almost doubled the level of recruitments to their sites by dropping delivery charges altogether. In the UK Asda charges £3.50 for delivery, against Tesco at £5.

04 July 2000

UK leads Europe

Nearly 70% of the leading dot.com companies are based in the UK, according to a new survey.

Six out of the top ten, and a total of 67 companies out of the top 100, figure in the list. Of the top 100, 29% are B2B while 41% are focused on the consumer. 36% of the founders have MBAs and a majority are in their early thirties.

The league was compiled by consultancy group Bathwick and published by the Sunday Times. No significant food companies make the list as yet, but wine features at 39 with Madaboutwine.com and at 55 with Wineandco.com [http://www.wineandco.com/]. The major food appearance is the European B2B site Foodmanager.com at 84.

04 July 2000

Big Mac pays online

MacDonald's revolutionizes the payroll.

McDonald's has junked its traditional payroll system in favour of a Net solution which has reduced time, cost and error in processing paychecks to the 110,000 US employees who work in its 1,900 corporately owned restaurants in America. Staff director Ginnie Cronin, said: 'We were faced with an old legacy system that actually featured three separate payroll systems. Every time a change occurred [in the payroll procedure], we had to make the change in three different systems. Now we have one centralized system handling all the payroll, so we're obviously saving money in maintenance and support'.

The new system works in real time, whereas before, employee cheques and benefits information were always handled in batches. Cronin said that tax deductions in particular have been handled more accurately and efficiently since the new system was adopted. For the next phase, workers will enter and update personal information that may affect their employment status or benefits.

04 July 2000

American giants merge

USA distributors sign up to B2B portal.

Global Food Exchange, the leading supply chain integration provider for the American food industry, and UniPro Foodservice have formed a strategic alliance to provide an e-commerce purchasing solution for UniPro's Member distributors.

UniPro is the leading co-operative of foodservice distributors worldwide with total purchasing power of over $20 billion and more than 200 distributors, including 22 of the top 50 distributors in the United States.

11 August 2000

Bean auction trial saves 15%

Net auction secures 100 bids.

A trial auction of pinto and navy beans yielded savings of 15% on the Net. The value of the transaction was £1.5 million. Ecfood.com [http://www.Ecfood.com] prepared the specifications and arranged for 10 suppliers to participate. Each supplier bid an average of nine times, for a total of nearly 100 bids.

'Our experience with these auctions has validated the strategic value of EcFood's services,' said Paul Dubendorf, vice president of finance at Furman Foods. 'This was our first time using the Ecfood methodology, and we were able to realize significant cost and time savings. We are now planning several additional transactions.'

11 August 2000

Wal-Mart signs up Shopsmart

Asda's parent company has signed a deal with the price comparison site ShopSmart in a move seen as a first step into the German market.

Wal-Mart together with AOL is expected to take a combined stake of just over 20 per cent in ShopSmart [http://www.ShopSmart.com]. The association of Wal-Mart with a popular price comparison service may reinforce the US group's claim to offer lower prices than many of its big supermarket rivals. Wal-Mart will inject its rival price comparison portal, Valuemad, into ShopSmart. Wal-Mart owns the UK's 250 Asda supermarkets, and 100 in Germany.

ShopSmart allows consumers to search 2,000 online stores for the best deal on a range of goods and services, such as concert tickets, holidays and clothes. It also offers reviews of the online stores. ShopSmart was the first website to introduce comparative shopping in the UK last year. This year it expanded into Germany and Sweden. It was ranked second in the Taylor Nelson dotcom weekly awareness survey, behind Freeserve, the internet service provider, although it has since fallen back sharply.

The group was founded in May 1998 as ShopGuide by Martin Reeves as an impartial guide to buying online in the UK. It derives its revenues mostly from advertising and by receiving a fee every time someone clicks on one of its links to an online shopping site. It does not charge stores to be on its site.

ShopSmart is planning a stock-market listing at some stage, though no date has been fixed. Analysts believe it would be worth more than £200 million.

14 August 2000

USA B2B portal sets up Korean venture

Agriculture giant Tradingproduce.com has moved outside of America with a deal in South East Asia.

TradingProduce.com [http://www.TradingProduce.com], the world's premiere business-to-business e-commerce site for the agriculture and food industries, has announced a joint-venture agreement with Tong Yang MoolSan Co., Ltd., which is a multi-billion dollar Korean conglomerate with a heavy emphasis in agriculture.

'This will be a full-blown operation with sales, marketing, engineering and a data centre in Korea,' said TradingProduce.com CEO Rob Bonavito on announcing the link up with Tong Yang MoolSan Co., the multi-faceted conglomerate based in Seoul.

14 August 2000

Big Mac sets up trading portal

An alliance of some of the biggest food service companies in America is to launch its own e-marketplace next year.

Cargill, Sysco, Tyson Foods, and McDonald's said they plan to streamline the food service industry's sales and purchasing practices by forming an independent marketplace. 'There are too many people dealing with the product to get it to the end consumer,' said Diane Sanders, vice-president and treasurer of food distributor Sysco.

The deal unites some of the major players in food service. In addition to McDonald's, the industry's biggest retailer, Sysco is the largest North American distributor of food service products; Tyson is the world's largest producer and processor of chicken; and Cargill processes and distributes food products in 60 countries.

14 August 2000

China sets up online food exchange

The Hong Kong Stock Exchange has backed a B2B portal for China State Farms Agribusiness with an eye on the $76 billion Chinese food market and predicts 30% of Chinese trade will be online within two years.

FoodNet100 brings with it a standardized, automated trading and logistics infrastructure to encourage economies of scale and streamlined processes in a secure and confidential digital environment. In addition to trading facilities, FoodNet100 also provides the most up-to-date information critical for the food industry, including news, market movements, price and food-related regulations.

'The food industry is one of the five largest in China, and receives the highest level of central government support for further development,' said Laurie Kan, President and CEO of i100 Limited, the technical company handling the exchange. 'In view of the large food market and the enormous number of buyers and sellers in China who are currently trading offline in a very fragmented and inefficient manner, FoodNet100 is obviously the most effective and efficient way for information exchange and food trading in the lead-up to China's expected entry to the World Trade Organization'.

China's agricultural products and food industry is one of the world's largest, comprising 130,000 food manufacturers supporting a population of 1.3 billion, generating $71 billion in annual revenue. In addition, 4,000 food producing and packing machinery plants generate US$5 billion in annual revenue. It is forecast that 30% of food-related enterprises in China will be conducting their business online by 2002.

Over 200 associate members under the Beijing Food Association will be enlisted as members of FoodNet100 and begin using FoodNet100 as their key channel for information exchange and trading in food products and food processing equipment.

'The food industry in China is burdened with entrenched inefficiencies,' said Kent Lam, CEO of FoodNet100. 'As the most efficient online food industry exchange, we help users to drastically reduce time spent in searching for critical information, to greatly reduce transactional costs and time spent in negotiating and executing each trade, to reach numerous local and foreign food markets at the touch of a button, and to facilitate the overall global development of this very important Chinese industry.'

The China State Farms Group is the largest central-government-owned food conglomerate in China, comprising of a total of 22 domestic companies and 16 overseas subsidiaries, with total assets amounting to US$570 million. China State Farms' business includes processing factories across three continents with more than 1,000 commodities and a global marketing network covering 60 countries and regions.

04 September 2000

Sweetener auction saves 6%

A collective auction, claimed to be the first of its kind, where different buyers pool their needs to make a single bid has saved six percent on prices across the board.

Multiple buyers from the dairy and canning industries were able to watch pricing changes within minutes as the auction transpired in Internet time. ecFood.com [http://www.ecFood.com], an independent trading web-base market, sourced the sweetener, qualified the suppliers, trained both the buyers and suppliers on online auction procedures, and oversaw the auction as it progressed.

'These types of auctions could ultimately change the dynamics of the market-place,' said Dave Laukat, CEO of ecFood.com. 'Conducting demand aggregate auctions over the Internet expands savings opportunities and gives automatic purchasing power to smaller and mid-sized manufacturers very similar to large purchasers.'

Forbes Magazine recently listed the company as 'Best of the Web' for B2B sites in the food industry.

04 September 2000

Efdex hits the rocks

The ambitious UK B2B portal and TV station efdex.com has called in the receivers.

The site had funding of £43 million which it invested in glossy brochures, exhibition spaces, TV plans and lavish headquarters near Reading. It had planned to be a trading exchange for more than 40,000 businesses but had only managed to secure agreements with 2,000. Some staff, of whom there were more than 200 in the Spring, claimed they had not been paid for months.

Efdex.com was one of the first companies to move on the B2B market, being set up in 1994, but despite first-mover advantage had been unable to secure core business.

05 September 2000

Kraft buys into EthnicGrocer.com

The deal was said to be a partnership to explore e-commerce data and opportunities to trade.

'EthnicGrocer.com [http://www.EthnicGrocer.com] is a leader in the procurement, distribution and online retailing of ethnic consumer products from across the globe using our proprietary technology and supply-chain platforms,' said Parry Singh, President and CEO of EthnicGrocer.com. 'Leveraging this expertise on top of Kraft's existing systems knowledge, EthnicGrocer.com is well positioned for future growth.'

'This e-partnership – along with others like Food.com [http://www.Food.com] – further expands our participation in online food solutions,' noted Paula Sneed, President of Kraft's E-Commerce Division. 'We will continue to explore joint strategic initiatives that offer synergies and growth potential for Kraft.'

About EthnicGrocer.com, Inc.

EthnicGrocer.com, based in Chicago, has engineered a technology-centered global supply chain that sources the largest selection of ethnic foods and products from around the world for consumers and businesses. The company developed and launched four web store fronts on top of its technology and supply-chain infrastructure: www.Namaste.com – targeting the Asian-Indian market; www.GongShee.com – serving the Chinese market; www.QueRico.com – targeting the Latin American market; and www.EthnicGrocer.com – marketing to cooking enthusiasts.

Kraft Foods is a leading e-commerce player with high sales of meal solutions off its kraftfoods.com [http://www.kraftfoods.com] site, according to PCData Online.

05 September 2000

B2B trail shows spectacular savings

More evidence of the joys of B2B trading as the Net helps companies work together.

A 6-month trail involving 12 manufacturers using B2B techniques showed a 4% growth in sales for retailers and a near 70% reduction in invoice deductions. The pilot allowed firms to synchronize pricing, promotions, sales, inventories, invoices, and other data via the Internet, and also nearly eliminated item and price mismatches. The scan-based trading allows retailers to pay for perishable produce at the cash desk not at the back door at delivery. Key findings were:

- Sales volume increased: sales in Schnucks' stores increased 3.9% compared to stores not participating in the pilot. Sales increases for suppliers ranged from 2.5% to 5.2%.
- Invoice deductions were reduced by 69% across all synchronized suppliers. Suppliers saved 20–25 minutes per delivery and retailers saved 10–15 minutes in delivery time.
- Time spent resolving item and price discrepancies was cut in half, from an average of 5 days at the pilot's outset, to 2–3 days at the pilot's conclusion.

'Because participants could access the data via the Internet, we were truly able to bridge the gap in technology capabilities between trading partners,' said Joe Hoff, Vice President, National Retail Sales, Anheuser-Busch Companies, Inc. and chairman of GMA's DSD Committee. 'The pilot also proved that smaller, independent retailers, like Andronico's, can implement scan-based trading without a significant investment in new e-commerce systems.'

Manufacturers in the scheme included Coors beer, Dr Pepper, Frito Lay and Kraft Pizza Company. For more information: Gmabrands.com [http://www.Gmabrands.com]

05 September 2000

Electronic boxes on the doorstep

A new UK trial in Reading is testing the viability of customised electronic boxes on the doorstep for secure home delivery.

Following on from USA developments like Zbox, Lantes and Brivo, a London company has developed Homeport which it is about to trial in 1000 homes around Reading. An electronic box is fitted outside the front door. The delivery operator swipes his smart card past the bar code and can then secure his box with a metal cable. Homeowners can then open the box with their smart card when they want and also place returns in it for the next delivery.

The developer Mark Lunn claims the system might increase the number of deliveries per hour from the current best performance of about 10 up to 15 – but also opens the possibility of night time delivery and cuts down wasted time in customers not being in when deliveries are made. Homeporthome.com [http://www.Homeporthome.com] has signed up specialist companies like Foodferry.co.uk [http://www.Foodferry.co.uk] and Berry Brothers wine merchants as well as gift and laundry companies. Trials are also scheduled for Lyons in France and in Germany. The advantage Homeport has over USA rivals – and also the Australian Eezeebox – is that it is tailored to European-style doorsteps where the American focus has been on larger chests for homes with plenty of porch space. Not that it will necessarily have the market to itself, as another company Dynamix has a similar scheme in prototype.

'Delivery is the single biggest issue facing retailers and e-tailers at the moment because it relies on people being at home,' said Lunn. 'This is unworkable because most people in today's world are simply too busy to hang around waiting for the doorbell to ring'.

11 September 2000

Interview: The man behind efoodmanager.com

B2B site Efoodmanager.com is one of the oldest established B2B sites in Europe and is building its UK proposition this autumn.

Geoff Jankowski is head of international operations and has set up the UK operation. Efoodmanager.com [http://www.Efoodmanager.com] started in fresh produce but has recently moved into fish and farming as well.

• How did efoodmanager come about?

'Back in '96 Fruchtmetz.net was set up by four people as a means of reducing the losses in distressed fresh produce in Germany. The question was what to do with the reject produce. Traditionally this was done by fax and would often be sold at a loss or for nothing. So we looked for a realtime solution. Initially it was part of Tengleman, the second biggest retailer of produce in Germany, who wanted to use it for their traders. But the company was separated in January because the importance of a trading environment is that it is genuinely independent, otherwise you do not have a free market.

• How has the UK been different to Germany?

'It is different. The market place is more integrated. And also perhaps because the UK was more advanced on the Net than Germany, people were prepared to sit back and wait and see where in Germany the need was to create a solution.

• How will efoodmanager.com develop?

'We have integrated into fishing with another site DerFischman and are currently looking to partner with banks so that we can invoice across countries. Over the next 12–18 months we will be developing in the UK, Spain, France and Italy and have just opened an office in Poland and have plans for the Middle East. From day one we have been sourcing globally anyway.

• What is the business model?

'We see a transactional model which will develop into a subscription service. There is no one model really because with the Internet we see it as important to test and try out different revenue streams. But we do see that value added services off the Net will be increasingly important.

• Such as?

'Well for example, if I am a German buyer purchasing beef from Argentina I can organise through the Web a personal visit to audit the goods and satisfy myself that the quality I am getting is OK. Also we can display multi-language solutions so that you can buy from each other without speaking the same language. The only thing you have to agree on is the currency.

• Who do you see as your main competitors?

'Well efdex.com (now in receivership) had very low grade technology. But we believe it will be easier for Europe to integrate as a market than America so perhaps sites like Globalfoodexchange in the States will have a different focus.

- Who do you see being the winners and losers with the arrival of e-commerce?

'The losers will be middle men who do not really add value, such as import and export agents. But the Internet will create new business. There are fantastic opportunities in the area of logistics for example.

- How quickly will all this happen?

'Well it has to be said that we are still explaining the benefits of the Net to companies and showing them how it can be used to integrate in existing businesses. But this kind of trading is not really that new. People have been using e-commerce in the petrol industry for years. What we are talking about is bringing the Internet tools into an industry that is not technologically literate. Also the food business – with the exception of companies like Macdonald's – is not vertically integrated and the Net will help to bring this about.

- Define efoodmanager.com's key benefit

'We are neutral. We are not aligned to any one company, we do not even have a stake holding or any investment money from anyone in the food sector. We believe this is important because otherwise companies might feel their trading information might be seen by competitors.'

11 September 2000

Produce B2B site moves to seamless logistics

Tradingproduce.com has set up automated transport and tracking software in a further move to seamless trading.

Buyers will be able to use the Net to tender, specify and deliver orders and allow transport companies to accumulate multiple orders. 'It eliminates the redundancy of re-entering the data for transportation companies,' said Zia Zahiri, Chief Technical Officer for TradingProduce.com [http://www.TradingProduce.com].

'Release II of our logistics is built on top of our well-proven transaction processing system. The two systems are tightly integrated to remove redundancies, improve accuracy of information and expedite the processing of the orders.'

Buyers, sellers and the transportation industry can download and upload reports, digital images, bar code information and bills of lading on-site. 'We are enabling technology to make the flow of information faster, easier and mobile. This is another step to eliminate the paperwork involved,' Zahiri said.

12 September 2000

Hewlett Packard invests $6.1 million in Foodmarketplace.com

An innovative online foodservice company has won the backing from the computer giant to develop radical new services.

FoodMarketplace.com [http://www.FoodMarketplace.com] and HP have been working together since June 1999 to develop Resourcelink, a program enabling food manufacturers to donate and relocate excess inventory to qualified charities over the Internet. To date, Resourcelink has facilitated the donation of more than £5 million. FoodMarketplace.com plans to launch SurplusFoods.com and FoodSamples.com later this month. 'SurplusFoods.com will allow food manufacturers to combine the sales and donation process into one continuous flow of information and decision making, while reducing inventory and transaction costs and preventing gray market sales.'

FoodSamples.com allows manufacturers to introduce new and featured products directly to restaurant operators. FoodMarketplace.com was founded in 1997 by Jody Helfend, Tim Plotkin and Charlie Jackson as a solution provider for surplus inventory. The Company has grown into a supply chain solution provider focused on the foodservice industry. The Company provides Internet-based solutions to more efficiently solve traditional marketing, selling, and donation challenges in the foodservice industry.

13 September 2000

California set up big wine play

B2B and B2C merger heralds start of major wine portal.

eVineyard has acquired WineBuyer.com [http://www.WineBuyer.com], the industry's first online B2B marketplace linking licensed alcohol beverage buyers and sellers. In August, eVineyard announced its first B2B play, with the acquisition of Restaurant Wine, a subscription-based newsletter targeted to the on-premise market with over 2,000 subscribers in the United States and abroad.

eVineyard's assertive expansion demonstrates the company's commitment to furthering its leadership position in Web-based wine retailing. John Boone, founder, chairman and CEO of WineBuyer.com, has been named vice-president of business services at eVineyard. Boone noted that the acquisition 'expands eVineyard's access to the wines its customers seek while giving distributors an unprecedented opportunity to benefit from the market access provided by eVineyard's virtual storefront. It's a perfect match.' WineBuyer.com, based in Palo Alto, California, was developed by seasoned technology and industry professionals to help automate and streamline B2B sales efficiencies between wine wholesalers and retailers.

eVineyard is the only multistate licensed online wine retailer and the leader in the B2C wine market, covering wine buyers in 25 states – more than 70% of the United States market for off-premise premium wine sales.

14 September 2000

Profile: Transora, the birth of a giant

Massive B2B play in America picks up 53 major players in packaged goods.

Chief Executive Officer Judy Sprieser has spent her first months in office persuading the regulatory bodies that Transora would be a service facility and not a cartel. Over the next 9 months, pilot trials will be underway to connect brands with the major retailers like Carrefour, Sears and Metro. The first stage is expected to be launched later this year concentrating on procurement and cataloguing and early next year financial management and forecasting are to be added.

The 53 member companies have invested according to turnover but none will be allowed more than five percent of the equity.

Sprieser, who joined Transora from Sara Lee's $8bn Food and Foodservice Group, is promising to cut the slack out of the supply chain by focussing on 'outmoded processes and the lack of industry-wide communication and co-operation'. Inflated procurement costs, excessive unwarranted administrative costs, large 'safety stock' inventory expenses, and expensive media purchases have been cited as main targets. The exchange's aim is to facilitate industry-wide supply chain integration and in the long term to provide innovative, problem-specific solutions.

Sprieser said: 'The consumer packaged goods industry spends trillions of dollars on safety stock just in case its predictions are wrong. Being able to reduce that investment in the supply chain alone will provide a lot of cash release.' She claims it will be possible to cut the costs of transactions from $150 to $20 by using the Net.

Sprieser said: 'We are different from the others (exchanges) who are owned by major technical application providers or venture capital companies. We know our business and its demands. No one will be able to approach our scale. We have over half the industry in our exchange. We will be a full service provider and will not exclude anyone. We have even got companies outside of our industry wishing to join us.

'It is going to be the companies which move fastest that will take advantage of e-commerce. They will be the winners.'

14 September 2000

But Nestle and Danone bid to keep Transora out of Europe

£30 million investment to create Euro rival.

Four of Europe's largest companies are setting up a B2B exchange in a bid to prevent US counterpart Transora from gaining too much of a lead in Europe. CPGmarket.com [http://www.CPGmarket.com] is spearheaded by Nestlé, the world's largest food producer, and also includes Danone, France's number one food group, which has a leading market presence in dairy food, biscuits and mineral water. The core non-food partners are the German chemicals group Henkel and SAP, the software provider.

Significantly, the US subsidiaries of both Nestlé and Danone are members of Transora. CPGmarket.com chief executive Yves Barbieux said it was essential that Transora had a competitor to ensure that the US cluster could not simply charge whatever it liked.

14 September 2000

Fish site lures the money with discount rates on transactions

Bucking the trend, American industry fish portal has dropped subscription charges in favour of a three-tier transaction charge on sales.

Gofish.com [http://www.Gofish.com] is one of the leading global seafood trading platforms selling a value-added, neutral exchange for subscribers.. The site features news but is also backed by a credit information bureau, Seafax, which can validate the credit worthiness of traders in real time.

Chief operations officer John Cummings revealed that transactions have been doubling each month since the site launched last November. Abandoning subscriptions, Cummings has introduced a three-tier commission system.

- Collection guarantee – where customers are assured payment if a buyer defaults at a cost of 0.75%.
- Receivables management – or an enhanced service which includes collection of the goods for 1% commission.
- Advance payment – sellers get up to 80% payment within 48 hours of the buyer accepting the product. Sellers pay 1% transaction fee plus the cost of money, which could equal a 1.7% fee.

Gofish is a privately held corporation based in Portland, Maine, with offices in Seattle, New York City, Los Angeles, Norway, Ecuador and Thailand.

21 September 2000

Foodtrader.com deals $15million in new funding

Chairman Frank Tomasino said the money would be used to develop full online trading.

'We have positioned ourselves to become the leading full-service solution for the entire food and agriculture industry. This financing gives us the necessary financial support to achieve that goal,' he said.

Andrew Russell, principal of East River Ventures, a major investor said: 'In today's environment, the only B2B companies that will succeed are those whose management teams possess in-depth knowledge to address industry-specific issues. Foodtrader has such a team.'

Foodtrader.com [http://www.Foodtrader.com] offers a marketplace exchange, value deals (excess inventory clearance), storefronts (online branding through transactional Web pages), reverse auctions for RFQs (request for quotes) and core logistical services, including trade finance, insurance, transportation and inspection.

Miami-based Foodtrader.com was recently named 'Best of the Web' by Forbes magazine. More than 13,000 buyers and sellers in 180 countries use the site to trade food products more efficiently, consult with industry experts, and take advantage of a complete global sourcing and purchasing solution. Foodtrader.com was the first online business-to-business food exchange to establish a presence in South America when it opened offices in Argentina in 1999. In 2000, the Company opened offices throughout Latin America and Europe to serve 33 international markets, giving it the most extensive worldwide presence of any US-based online food exchange.

17 October 2000

Irish meat company develops full traceability software

A meat marketing company has developed software together with a group of organic farmers in Shannon that allows for full traceability back to the farm on all animals on any product.

This revolutionary new way of working also allows farmers to identify what happens to their meat further down the food chain and at the same time allows consumers to know exactly where their food comes from. Director of Greenstuff, Michael Barsham said: ' We go the last mile. At present supermarkets can trace back where they get their foods from, but consumers cannot. We use the system to follow the ear tag process right from the farm through to the label on an end product like sausages.'

Greenstuff is an organic beef marketing company, which represents 85 farmers in Shannon. 'Farmers told us they had no idea where their meat ended up, so with this system we can give them feedback right from the customer end, so they know if their products are well received or not. It also helps us to know when the farm will have animals ready for market and we can help the farmers schedule their income streams,' said Barsham. Contact: greenstuff.org.uk [http://www.greenstuff.org.uk]

18 October 2000

One billion litres of wine sold on the Net in the last four months

Worldwinetrade.com is claiming to have traded more than a billion litres of bulk wine on its site in the last four months.

The better-than-expected transaction volume is due in part to a membership of leading industry players from 18 countries around the globe and more than 41 million litres of bulk wine listed online. Members in South America, South Africa, and Europe have successfully used WorldWineTrade.com [http://www.WorldWineTrade.com] to sell wine to bottlers in Europe, North America, and Scandinavia. In one instance, a well-known California winery used WorldWineTrade.com to purchase hundreds of thousands of litres of Sauvignon Blanc from three different Chilean producers, resulting in savings of 10-15 percent on wine and shipping costs compared to traditional offline purchasing methods.

Mike MacDougal, Chief Executive of the parent company BulkMarkets.com [http://www.BulkMarkets.com] said: 'The early success of WorldWineTrade.com proves that specialty bulk commodities industries are ready for B2B e-commerce. Bulk wine buyers and sellers appreciate the cost and time savings provided by WorldWineTrade.com, and BulkMarkets.com will soon meet the demands of the juice concentrate industry with the upcoming launch of our next vertical, WorldJuiceTrade.com (www.WorldJuiceTrade.com).'

BulkMarkets.com is generating revenues by providing industry-specific channels where producers and packagers execute their entire transaction online aided by real-time market data and news about pricing, volume, and product availability.

'Companies eager to streamline business practices look for one place to fit all their needs. With online content available prior to transactions as well as shipping and logistical services offered upon completion, BulkMarkets.com's e-marketplaces are the one-stop solution,' said MacDougal.

BulkMarkets.com is also poised to scale its patent-pending technology into other specialty agricultural bulk commodity markets including nuts, dried fruits, spices/herbs, and edible oils.

BulkMarkets.com completed its first round of institutional funding in mid-March with investments from Bay City Capital's North American Nutrition and Agribusiness Fund, William Blair New World Ventures and William Blair & Company.

18 October 2000

Wine portal buys into negotiative software

B2Bwine.com has set up online negotiating tools to allow importers, distributors and producers to trade online.

It is calling its new RF software 'a proprietary dynamic negotiation platform'. It will allow qualified buyers to securely and privately post current and future product requirements, indicate price and volume information and request additional information and quotes from B2Bwine.com [http://www.B2Bwine.com] member wineries. Member wineries will use RFwine to search for sales opportunities throughout the world and begin to negotiate terms with buyers and importers.

'The new RFwine system will allow wineries to stay abreast of buyer requirements throughout the world, and this is a much welcomed benefit for us all,' said Christian Sotomayor, International Business Director of Vina Santa Ines & De Martino.

'We have spent the past nine months working to ensure that RFwine meets the needs of B2Bwine.com's members,' said Matthew Cole, CEO of B2Bwine.com. 'RFwine will launch in January 2001 with extensive input from B2Bwine.com's 500 member wineries, importers, distributors and buyers.'

'We are tightly focused on providing value-added solutions to our members without extracting commissions,' commented Arthur Massolo, President of B2Bwine.com. 'The combination of a sustainable membership-driven revenue model and intelligent focus on maintaining a low-cost structure will allow B2Bwine.com to achieve profitability during the first quarter of 2001.'

B2Bwine.com launched in January 2001 and has more than 500 members from 23 countries. More than 350 leading wineries and 150 international buyers, importers, and distributors have begun using B2Bwine.com to search out new contacts, request product samples, and negotiate price and volume terms.

18 October 2000

---

ecfood.com launches new suite of online tools for purchasing

The headlining gimmick is its ecProcure Negotiated Buy function which claims to advance online buying past the point of simply lowest price.

These solutions were created to manage the complex procurement process of the industrial food sector. ecProcure, Negotiated Buy, Supplier Rating System, Supplier Profiling, Specification Management and Professional Services are among the family of ecFoodware online tools based on state-of-the art technology.

'Our ecFoodware provides buying solutions based on real-world food pur chasing,' said Dave Laukat, CEO of ecFood.com [http://www.ecFood.com] . 'In particular, Negotiated Buy with its Supplier Rating System and Supplier Profiling takes the 'best practices' of industrial food purchasing and blends them with the efficiencies and savings of the new economy. Our services give tremendous control to both buyers and sellers to negotiate based on complete ingredient specs and purchasing goals beyond price.'

At the heart of ecProcure is the purchaser buying function. ecProcure offers three forms of buying strategies. One strategy is the standard, dynamic pricing auction. This is a private competitive auction, conducted in Internet time. Privacy is ensured through restricted access to invited participants.

Negotiated Buy is the newest and most sophisticated ecProcure purchasing strategy with the Supplier Rating System (SRS) feature. The buyer uses the SRS to rate suppliers according to specific criteria. The buyer determines the item, clearly defines the purchasing criteria, and rates their suppliers. During the bidding procedure, the buyer and supplier negotiate online to meet the price and criteria that was initially defined. The SRS, which is customised by the buyer and can include wide ranging criteria, allows buyers to better assess the value of a purchase beyond price. It also provides clear communication to suppliers.

There is also a free form option where the buyer can define the requirements and negotiate with multiple sellers on whatever purchasing point they want.

Additional ecFoodware solutions include ecProcure demand aggregation, and storefronts, specification management, supplier profiling and professional services.

ecFood.com is based in Santa Clara, California. Forbes Magazine recently recognised ecFood.com as 'Best of the Web' for the B2B sites in the food industry.

18 October 2000

Foodvision.com forms B2B alliance in restaurant kitchen sector

Further consolidation of the online market in America has seen Foodvision.com's B2B site form an alliance with Abestkitchen.com.

AbestKitchen.com [http://www.AbestKitchen.com] has been providing commercial quality supplies such as grills, fryers, ice machines and bar stools to restaurants, bars, and food service operations throughout Ohio (their homeland), the mid-west, and across America. It will provide Foodvision.com's independent restaurant operators with top quality restaurant and household products. They will also be able to take advantage of placing orders 24 hours a day, and have access to an extensive catalog of products.

'Foodvision.com [http://www.Foodvision.com] is clearly on its way to offering the only site meeting all the business needs of the approximately 450,000 Independent Restaurant Operators in North America,' said Paul R. Smith, CEO of Foodvision.com. 'We anticipate many strategic alliances by this time next year, adding millions of dollars to our revenues.'

AbestKitchen.com was launched in May 1998. The company's Internet store enables professional food establishments and consumers worldwide to order equipment and supplies online. Abestkitchen.com also features give-away drawings, a directory of resources, message boards, and gift services.

'Foodvision provides a valuable resource for food service professionals and food enthusiasts. Their leadership and innovation in online food makes them an ideal partner for Abestkitchen.com. We are very excited to be working with the Foodvision.com team,' states Rick Papaleo VP, Head of Internet Strategy for AbestKitchen.com.

www.Gate2Plate.com [http://www.Gate2Plate.com] is a B2B e-commerce marketplace spoke of Foodvision's site for the independent restaurant operator in North America. The site provides all the information an independent restaurant operator needs to effectively operate their business and the opportunity to save money through participating in an extensive e-commerce marketplace.

Foodvision.com was launched in November 1999. The site is targeted to the general public interested in all food-related topics and food service professionals interested in all food-business-related issues.

Foodvision.com includes hot topics, recipes of the day and week; restaurant reviews, creating interesting menus, restaurant design and financing, cooking schools, etc. Foodvision.com, Inc., is a food-specific Internet information provider and portal company that integrates traditional business systems in food and beverage operations with leading-edge technologies in Internet systems.

18 October 2000

Now even sweets go for B2B

It's wrapped, its sticky, its online . . . the World's first candy marketplace.

eCandy, the first full-service Internet destination for the confectionery community, has launched eCandy Marketplace, a business-to-business website at www.eCandy.com [http://www.eCandy.com].

eCandy Marketplace will serve as the premier destination for manufacturers, distributors, brokers, and retailers to source products, gather information and interact with each other on a secure website. The company's consumer-focused website remains at www.candy.com [http://www.candy.com].

Registered members can see which products are truly in demand; read exclusive industry news provided by Stagnito Communications, publisher of Candy Industry and Confectioner magazines; track purchases and sales; and check account and transaction history. A comprehensive site tutorial will help first-time users navigate the website.

James Griffin, eCandy President, said the site would kick start a 'confectionery renaissance'.

The launch of eCandy Marketplace comes on the heels of securing an undisclosed amount in second-round funding led by Crescent Venture Investors. The round, which included investors Staenberg Venture Partners and T.I. Capital, was completed last month.

Founded by Rani Aliahmad, John Hadl and David Kim, eCandy is based in Los Angeles, and is a privately held company whose investors include Crescent Venture Investors, T.I. Capital, Staenberg Private Capital, and iXL Ventures.

20 October 2000

Now smart companies post their results on line

And the executives web cast their stories live on a webcam.

The latest company to jump in is Sara Lee which put up its results in October and vice-chairman Janet Bergman, Sara Lee's vice-president, investor relations and corporate affairs webcast for a whole hour.

20 October 2000

Wine gets groggy on too many portals

The number of wine portals grows seemingly by the day to the point that B2B options runneth over.

Winebusiness.com [http://www.Winebusiness.com] is claiming now to have commissioned EC Cubed 'to create the wine industry's most advanced B2B eMarketplace'.

'Unlike off-the-shelf solutions, EC Cubed's build-to-order component-based software platform can be continually adapted to meet the unique needs of the wine industry,' said Richard Shell, CEO and founder of Winebusiness.com. 'Our market changes constantly with new technology, evolving demands and the seasonal and year-to-year fluctuations of any agriculture-based industry,' he said. 'EC Cubed is the perfect match for us because it can provide an eMarketplace with built-in flexibility.'

'Winebusiness.com is building an eMarketplace to meet the unique demands of the wine industry. They wanted far more than a packaged application could provide,' said Mike Teeters, senior vice-president of sales and marketing at EC Cubed. 'Our dynamic approach to building eMarketplaces will provide Winebusiness.com with the precise functionality they need in less time and at lower cost than a packaged solution. In addition, it will enable us to rapidly evolve their eMarketplace to respond to their changing business needs.'

Winebusiness.com was founded in 1992 as a trade publication and is headquartered in Sonoma, California, with offices in San Francisco.

Also in Sonoma is Vintners.com [http://www.Vintners.com] which claims to provide the only dedicated search engine for wine, spirits, and the beer industry on the Internet. Its Global Directory to wine, spirits, and beer is, it says, 'an indispensable marketing and product development tool for industry professionals, as well as the general public'.

Vintners.com Press Room publishes daily news releases from wineries, breweries, and distilleries from around the world, as well as from leading retailers and wholesalers for the beverage industry.

Vintners.com is now offering free online listings to all legitimate beverage industry companies. Members can create their own complete, interactive business profile.

20 October 2000

PlanetRice takes on funding

A global site for the rice industry has moved swiftly through the Internet chaos of the summer to secure funding to create a dynamic new trading platform.

Al Brooks launched the site after 20 years in the business and will be funded by Internet.com Ventures who have similar vertical content-led models.

Rice accounts for more than half the daily food of one-third of the World's population and is the most widely grown food grain. Rice is an important commercial export crop for Thailand, Vietnam, China, India, Japan and the United States. Extensive rice research around the world has doubled field yields and provided major improvements in quality and farming technology in the past 30 years.

Brooks said, 'We're proud to be able to provide online resources, technology, community and content for the rice industry, not only for the benefits that companies and organizations derive, but also for the benefits gained through strengthened international ties, a stronger sense of community and common understanding of the complexities and concerns that relate to this very large and important industry.'

Based in Magnolia, PlanetRice launched in December 1999 through word of mouth, direct email contacts, limited print advertising and search engine placement. Traffic has grown over 30% per month, from 6,000 page views in January to over 35,000 in November 2000. The directory contains well over 300 listings from every segment of the industry. The links directory contains dozens of rice-specific links. Planet Rice receives daily traffic from searches on all of the major search engines and is linked on all of the major rice industry destination sites as well as on many agricultural sites that carry links. Companies can self-publish, from their browsers, their own dynamic website within the planetrice.net directory.

Planet Rice, Inc. has signed a co-branding agreement with Sortex Ltd., a UK-based corporation whose brand is widely known throughout the rice industry.

Rice represents over 70% of the nutritional intake of 30% of the world population. 560 million metric tons of rice are produced each year. 4.7% of world production is exported from the top rice producing countries. The export market alone has a value (rice cost only) of over $3 billion.

17 November 2000

Thailand goes for vertical B2B site

Thai buyers and sellers have set up a global B2B site to trade online.

'We've created a true international marketplace where buyers and sellers of the top categories of food commodities come together and do business,' said Simon Chan, Executive Vice-President of FoodMarketExchange.com. 'We're offering products from the top Southeast Asian suppliers of seafood, grains, fruit and poultry to buyers the world over.'

FoodMarketExchange.com has signed 50 suppliers of foodstuffs, who represent a combined $2.5 billion (US) worth of exports last year worldwide. Major suppliers include Thai Union, TIPCO, STC and Surapon Foods. The company is now targeting the import buyers community given its unique position as the leading supplier of Southeast Asian food products.

Registered users buy by auction or fixed price. The auction feature includes both a traditional auction, in which multiple buyers bid on product from a single seller either in real time or via sealed bids, as well as a reverse auction in which a single buyer entertains bids from multiple sellers. The exchange enables trading between multiple buyers and sellers for one-time or standing orders.

The site also features a data center, which includes extensive news and information on market conditions, financial data, laws and regulations, unique services and specific product information. The site's 'People' section offers expert advice and opinions on current issues in food importing, as well as information on education and nutrition.

17 November 2000

Tesco hooks its check-out tills to the Net

Tesco has signed a three-year deal to link its cash registers through the Net for instant accounting.

The deal involves the Nasdaq-listed Point of Sale's global asp technology. Tesco will use some of the same functionality available via Point of Sale's US-based ASP, StoreAlliance.com, and its Israeli-based counterpart StoreAlliance.co.il, while relying on its own resources for IT support and infrastructure maintenance to the system.

The three-year agreement allows Tesco to deploy the ASP/thin-store concept initially in more than 800 stores in the UK, Ireland and France, and later in additional Tesco stores throughout Asia Pacific and Central Europe. The agreement also includes a corporate licence for Point of Sale's soon to-be-released checkout system which runs on the Linux operating system. The Linux-based solution will support Tesco's global IT strategy for implementing and maintaining a single point-of-sale system across all its 20,000 checkout lanes in eight countries.

The internally hosted, web-enabled system uses the Microsoft Internet Explorer web browser to provide Tesco store and chain managers with 24-hour, on-line access to operational information and store-specifics, such as transaction-level data collection, electronic journal and cross-reference sales reporting.

CEO Barry Shaked said: 'Our initial implementation of the web-based solution was focused on ASP and marketplace initiatives, with the intention of bringing leading-edge technology benefits to small chains and independent grocers. Tesco's decision to adopt this web-enabled software system is evidence that our new thin-store strategy is also suitable for a world-leader in the food retailing industry.'

Point of Sale Limited, with headquarters in Israel, provides integrated enterprise-wide software solutions for the retail food industry worldwide, including supermarkets, convenience stores and restaurants. With installations in more than 15,000 stores across 41 countries, the company markets its software solutions through direct sales, distributors, local dealers and through its US subsidiary, StorePoint. More information: www.pos.co.il [http://www.pos.co.il] or www.storepoint.com [http://www.storepoint.com] or www.storealliance.com [http://www.storealliance.com].

17 November 2000

Agribuys opens London office

Global food buying portal sets up for European business.

'This is the first in a series of moves designed to make Agribuys' technology and solutions available to major retailers, food-service operators and suppliers in the European Union,' said Bob Atkin, Agribuys' representative in charge of international partnerships. 'We have been encouraged by the enthusiastic response of the European market. Both food buyers and suppliers seem pleased with how well the functionality of our Web-based technology mirrors the activities of their relationship-driven supply-chain operations.'

Dhruba Kalita, Agribuys' executive vice-president and chief technology officer, continued: 'Agribuys' rapid global expansion has been made possible by two factors. Our technology platform was designed to accommodate multiple currencies, languages and time zones from day one; and we have just received $32 million in support from business-to-business e-commerce company Internet Capital Group (Nasdaq: ICGE – news) and our global strategy partners – and now have the necessary resources in place.'

Agribuys is a leading global provider of integrated Web-based procurement services for the food industry. Having recognized that the process extends far beyond a simple transaction model, the company provides preorder, transaction execution and post-order services in multiple food segments, including produce, meat, poultry, seafood, ingredients, dairy and floral. These services are seamlessly incorporated into its customers' front-end sales systems and back-end information systems, facilitating collaborative supply-chain management.

Based in California, with offices in India, Australia, Canada and the United Kingdom, Agribuys is rapidly expanding – more information at www.agribuys.com [http://www.agribuys.com].

17 November 2000

All purpose B2B seeks to rationalize market supply

A new B2B portal is about to launch in America aimed at bringing the fragmented food and drink sector into an online trading solution.

The Novopoint marketplace will give buyers and sellers of food ingredients, packaging and related services a single place to connect with each other, conduct transactions and better manage their supply chains. The company will serve participants of all sizes from every facet of the food industry, including buyers and suppliers of oil, sugar, colorings, packaging, chemicals, freight and everything in between.

It will be run by an independent company funded by the food and drink companies and powered by Ariba.

'The food manufacturing supply chain has numerous inefficiencies that make it ideally suited to the global economies of scale and process efficiencies of a B2B marketplace,' said Ariba President and COO Larry Mueller. The marketplace will offer supplier enablement, hosted and enterprise B2B procurement for goods and services, auctions, reverse auctions, bid/ask exchanges, strategic sourcing, spot buying, customer-specific pricing, electronic payments, logistics and online community forums.

The company argues that the $300 billion American food market is marked by buyer fragmentation among many manufacturers and incompatible, labour-intensive logistics and communications systems. Food ingredient purchases typically involve multiple, redundant sets of fax and phone communications throughout the purchasing, planning, shipping, tracking, receiving, invoicing and payment processes. Smaller food manufacturers also have limited access to a broad set of ingredients because it is not cost-efficient for large suppliers to transact with them directly.

Key partners are Cargill (www.cargill.com [http://www.cargill.com]) headquartered in Minneapolis, an international marketer, processor and distributor of agricultural, food, financial and industrial products and services with some 82,000 employees in more than in 59 countries. And California-based Crosspoint Venture Partners (www.cpvp.com [http://www.cpvp.com]), a leading venture capital firm specializing in B2B, with more than 50 B2B projects under way. Crosspoint portfolio companies, which went public in 1999, now represent market capitalization in excess of $100 billion.

30 November 2000

Americans move on the Spanish speaking market

A dedicated B2B portal for Spanish speaking companies is being set up to link South America with Spain and Portugal.

Initially aimed at food service using a template from America, the new portal brings together the Spanish phone giant Telefónica B2B, a leading B2B Net Market Maker, and B2Bgalaxy.com which runs Foodgalaxy.com in the USA. It is projecting savings of 20 per cent for buyers and 15% for sales.

30 November 2000

Net opens up Chinese market

Corporation acquires distribution along with online trading portal.

China Premium Food Corporation will combine the established food distribution business of Mandarin Fine Foods with its B2B e-commerce portal. B2B e-commerce this year in Asia is expected to hit nearly $30 billion, according to The Gartner Group, and could top $1 trillion in 2004.

Mr. Roy Warren, CEO of China Premium, said, 'To our knowledge, this is the first B2B portal launched that is specifically focused on bringing food products to China. We have very strategically followed a business model that has first established a firm bricks and mortar presence in major metropolitan areas of China prior to the launch of the B2B portal. We felt that it was too premature to launch the B2B platform earlier, without the immediate means to service the demand that we expect from the portal. We feel that the B2B platform will offer a streamlined process for Chinese institutions, hotels and the like, to efficiently procure premium-quality foods and makes available to US food producers a clear-cut route and distribution system in China.'

30 November 2000

Agribuys takes in $32 million

Leading world farm site gets big backing.

Internet Capital Group has injected $24 million of the $32 million total funding for B2B portal Agribuys to help it attack the $4 trillion global food industry. 'The food industry is ripe for a B2B e-commerce solution to streamline the complex process of bringing millions of food products from the farm to dinner table. The potential savings for the industry are in excess of $9 billion in the United States alone,' said Ron Hovsepian, managing director at ICG. 'Agribuys is a company that understands the complexity of this process, and has developed a solution that manages the intricate workflow of the industry.'

Agribuys' integrated procurement platform allows customers to procure and manage the supply chain for all food purchases including produce, meat, poultry, seafood, dairy, floral, ingredients and frozen perishables through a network of over 1,000 suppliers in the Agribuys marketplace.

'Agribuys is a powerful addition to the ICG partner company network in the food industry, which also includes eMerge, Cybercrop, FOL Networks, and Tibersoft,' added Hovsepian. 'With our stake in Agribuys, we are connecting every link in the food chain including farmers, ranchers, processors, manufacturers, grocers, distributors and restaurants. Based in California, with offices in India, Canada, England, and Australia, Agribuys is rapidly expanding to further broaden its global procurement platform.

30 November 2000

Nestle ties up Thailand

Nestle has announced a B2B agreement with Freemarkets.com to develop a sourcing portal for Asia.

Talley Goodson, vice-president and managing director of FreeMarkets' Asia operations. 'Nestle is a leading and innovative food company that joins a growing number of Asian companies who are leveraging the power of Free-Markets' technology, information and services to reduce their costs and sourcing process time and to make better sourcing decisions.

FreeMarkets has executed over 6,800 online auctions for more than $10.5 billion worth of goods and services to date, and created estimated savings of nearly $2 billion for its customers. The FreeMarkets B2B eMarketplace enables buying organizations to source products from more than 150 supply verticals. More than 7,000 suppliers from over 55 countries have participated in the FreeMarkets B2B eMarketplace.

30 November 2000

Farmers combine to harvest online advertising

The $500 million online farm supply industry is looking to further increase revenues by combining its advertising approach.

Engage Business Media is to create the industry's largest and most in-depth agriculture vertical audience group as part of its B2B online advertising network. With brand name sites such as Dairy Business.com, Farm Equipment Auctions and GrainFarmer.com, Engage Business Media provides the most valuable B2B agriculture ad inventory to online marketers.

Research indicates that the agriculture market is one of the hottest vertical industries on the Internet. The Red Herring reported that this year $500 million worth of farm supplies are expected to be transacted over the Internet, while Goldman Sachs reports that by 2004, US agriculture e-commerce is expected to reach more than $120 billion with an annual growth of approximately $20 billion.

30 November 2000

Asian B2B conducts global deals

A new agricultural site has announced that it is hosting deals as far away as selling vegetables from South America to North America and corn from Brazil to the Middle East.

After less than five months of operations, FreshfromtheWeb.com [http://www.FreshfromtheWeb.com] is claiming 725 registered members from around the world, all of them agricultural industry players. Of these 396 are in Asia, mostly Thailand, China and India, more than 125 members from Europe and 118 in North and South America.

Three business deals have been successfully concluded and fulfilled involving the sale of sesame seeds from Asia to the Middle East, the sale of vegetables from South America to the United States and the sale of yellow corn from Brazil to a country in the Middle East.

The FreshfromtheWeb.com currently offers 250 trading offers posted online and has announced a web-based spin off for IT services in Freshanswer.com using IT teams in Bangkok and India.

18 December 2000

Now recipes merge with coupons online

Marketing and recipe data site meals.com has announced a tie-in with coupons.com to give online cooks discounts for using different ingredients.

'People visiting www.Coupons.com [http://www.coupons.com] can now find great dinner ideas that use the money-saving coupons available on the Website,' said John Melin, vice-president of business development for Meals.com [http://www.Meals.com]. 'In addition to saving money on groceries, Coupons.com users will have the inspiration and tools to plan and shop for better meals.'

In addition to Coupons.com, Meals.com has agreements with a growing list of Internet companies; with food manufacturers like Nestle, Smucker, and Campbell; and with a number of supermarket retailers that operate more than 2,400 stores. Meals.com also powers in-store technologies such as the Shopper, a Web access terminal that provides customers with personalized offers and relevant information. It connects consumers, grocery stores and manufacturers through online and in-store technologies that provide shoppers with relevant information and special offers based on their personal preferences. Meals.com is also a subsidiary of Coinstar Inc, a coin-counting and e-services company that has relationships with more than 8,000 US supermarkets.

18 December 2000

Globalnetexchange claims $300 million of trade a month

One of the biggest global B2B plays in the food sector is claiming substantial trade within eight months of launching.

GlobalNetXchange (GNX), the leading global business-to-business online marketplace for the retail industry, reported that members on the exchange have conducted transactions in excess of $300 million through October. GNX is a consortium of worldwide retailers who have used the platform to trade clothes, electronic goods, cleaning and travel. Total purchasing power of members on the exchange is claimed to be $250 billion.

'The volume on the exchange in the past month has been double what we experienced in our first months of operation,' said Joe Laughlin, GlobalNetXchange CEO. 'As the savings and efficiencies to be realised from the GNX services become clearly proven, our retail industry partners are rapidly moving from a testing phase to integrating these tools into their business processes.'

Powered by Oracle Exchange, Oracle's online e-business marketplace is used by: Carrefour, SA, the largest retailer in Europe and Latin America and the first international retailer in Asia; Kroger Co., the largest US retail grocery chain; Metro AG, based in Germany and one of the world's largest retailers; J. Sainsbury Plc, one of the world's leading retailers and the United Kingdom's second largest food retailer; Coles Myer, the leading retailer in Australia; Pinault-Printemps-Redoute SA, the number one non-food retailer in Europe; and Sears, Roebuck and Co., a leading US retailer. More information – www.gnx.com [http://www.gnx.com].

18 December 2000

Sainsbury's ties up media play with Carlton TV

TV and Internet channel to be set up on a 50:50 play.

The Taste Network will bring together Carlton's SimplyFood website, the Carlton Food Network television channel and Sainsbury's food and drink heritage from its 'tasteforlife' website. The two websites will merge to form a new website, taste.co.uk, and a new television channel, 'tasteTV', which will be launched in spring 2001. Forty staff will also transfer to the new company from Sainsbury's and Carlton.

Taste Network will enable viewers to watch their favourite cooking shows, download the latest recipes, check out new restaurants and order food and drink on television and the internet. Anthony Ellis has been appointed Chief Executive of the new company. He said: 'Taste Network represents a unique partnership, bringing the very best brands together in television, Web, food and drink. We will be merging Carlton's strength in multimedia entertainment with Sainsbury's unrivalled food and drink knowledge.'

'Carlton has content, production and interactive expertise, while Sainsbury's has a nationwide retail presence, a strong e-tail business and distribution and fulfillment experience.' Patrick McHugh, Sainsbury's Group Director of e-commerce, said: 'This alliance is another step in the roll-out of Sainsbury's e-commerce strategy and recognises the importance of new media in e-commerce. The new venture will bring to life the promise of digital interactive television and enable viewers to place orders for their home grocery deliveries directly from their living rooms.'

Sainsbury's believes that the UK online grocery market will be worth around £7.5 billion by 2005. The Henley Research Centre suggests that shopping via the television will overtake internet shopping by 2004 and that over 90 per cent of UK homes will have interactive shopping via digital television by 2010.

18 December 2000

Big Mac creates its own B2B portal

Big Mac is to launch its own B2B portal which will allow worldwide franchisees to buy uniforms and services.

In a move that could profoundly alter the multibillion-dollar fast-food industry, a new Internet procurement site designed for McDonald Corporation's 27,000 franchises is almost ready to serve. The site will be launched by eMac Digital, a company founded in June 2000 and jointly owned by McDonald's and Accel-KKR Internet Co. The as-yet-unnamed procurement hub is the first to target the fast food restaurant industry, which is among America's largest and most profitable. The most ambitious part of eMac Digital's hub is its plan to include other franchises from other fast-food companies. The hub, set to go live in mid-2001, will allow all of McDonald's franchises across the globe to buy everything needed to run their restaurants – from uniforms to spatulas to hamburger buns – all on one site.

18 December 2000

Gofish adds news tools

Seafish B2B portal develops new online services.

Gofish.com has developed two new online tools to help buyers and sellers to trade: Saved Orders and My Markets. 'We've been listening to what our customers are telling us,' said Cheryl Bascomb, Gofish director of marketing. 'And because many say they place the same order every Tuesday or view the same markets each morning, we're adding Saved Orders and My Markets to make doing business easier and more efficient for our members.'

'The Saved Orders feature helps automate a routine order for our members, leaving them with more time to focus on the other, more difficult aspects of their business,' says John Snyder, Gofish species exchange manager. For example, a buyer may frequently buy a few thousand pounds of black tiger shrimp from a particular seller. Now, when the transaction is complete, the buyer can use Saved Orders to save that order, needing only to click on it the next time they wish to make the same purchase.

Gofish will also be adding a feature whereby members who place orders on a regular schedule can be emailed an order confirmation. If, for instance, a buyer orders salmon every Monday, they would receive an email Monday morning with their saved order asking them to confirm the purchase and/or update the information.

The second new tool, My Markets, will allow members who view the same market information on a regular basis to use that view anytime by clicking on My Gofish. So if a member frequently looks for information on yellow-fin tuna steaks, they can perform that search by clicking on My Gofish, instantly seeing pricing information and the number of buyer/seller listings and auctions. Customers can also click on the new View Market Snapshot to see all the listings on the Gofish site, sorted by the criteria that works for them.

Gofish.com is the leading global seafood trading platform and risk management information source. It provides a value-added, neutral exchange destination for industry buyers and sellers. The site features up-to-the-minute stories from the Gofish news team and the Seafax Business Wire, plus market information from industry-leading SeaFood Business. The portal also contains secure access to proprietary credit information on trading partners provided by the industry's most widely respected credit information bureau, Seafax. Subscribers to the exchange are guaranteed to receive the highest quality product and accelerated payment through the industry's only seller collection guarantee and receivables funding programs. Gofish is a privately held corporation based in Portland, Maine, with offices in Seattle, New York, Norway, Ecuador and Thailand. More information – www.gofish.com [http://www.gofish.com].

18 December 2000

Part 5
The Big Players

The corporate vision and application of e commerce for large scale companies such as supermarkets and major manufacturers is a sector apart. The European market is the goal, the massive logistical implications the obstacle.

Anger as retailers pass on IT costs to suppliers

Moves by large supermarkets to get suppliers to pay to buy into Intranet systems have been met with strong resistance.

Sums as large as £10 000, plus fees of £250 a month, have been demanded by retail chains, Tesco among them, to buy into their own Intranet systems. Many small producers have refused point blank to pay for the privilege. One supplier said: 'It is outrageous. We told them to get lost. There is no denying that the Intranet is a useful service with everything from weather charts to store contacts and right down to daily reports on sales, but we are just not prepared to pay. That is not proper business'.

The move might be seen as a predictable extension of the historic and unloved retail support budgets but also raised much wider issues as to who is actually pulling the strings in the e-commerce universe. For a mainstream manufacturer selling mostly through bricks and mortar stores, the up-to-date information probably has a much higher value than for a new supplier or a supplier looking to develop e-commerce. But if retailers are prepared to pass this level of detailed management information down the supply chain, the potential impact on traditional management jobs at retailer headquarters may be significant. As one insider pointed out: 'Total devolution of management responsibility could be just around the corner, but if a retailer is prepared to give up this area of responsibility then questions are soon going to be asked as to what the retailer is doing in the first place?'

To date Tesco is the most implicated in these demands plus some smaller retailers. Sainsbury, Waitrose and Safeway are still maintaining completely free use of their Intranet systems.

04 January 2000

Hazlewood draws the line on e-commerce

One of the biggest and most powerful suppliers to supermarkets has ruled out a move into e-commerce.

Ready meals giant Hazlewood [http://www.hazlewoodfoods.com/] has stopped short of a move into e-commerce and decided to use the web only for communications and information for potential customers.

Director of communications Jonathan Grant-Nicholas said: 'Our computer links are in essence direct and we are not currently aiming at e-commerce in the traditional sense. Our website is being developed as a source of reference to these major customers, our suppliers, agencies etc. in support of our position as one of the leading suppliers in our chosen areas. Our technical site, for instance, is referenced by our customers, suppliers and regulatory agents such as environmental health officers. As we develop we look for the site to become an information source for both outside reference and for our employees – in this respect the Intranet and Web are being developed to be seamless'.

06 January 2000

7000 new e-jobs at Tesco

Tesco has thrown its full weight behind Internet shopping committing 300 stores to e-tailing covering 90% of the country by the end of the year.

Tesco [http://www.tesco.co.uk] said that it now has 250,000 registered online customers, which it claims makes it the biggest Internet grocery retailer in the world. Under new plans the target is to get a million.

The supermarket says it will also expand into non-food online services, such as sales of books, clothing, gifts, furnishing and banking during the course of the year – turning itself into the biggest retail operation of any kind in the UK.

This will throw it into competition with the giant names such as Amazon.com in the book and music retailing sector: many of the High Street banks (it already has an alliance with the Royal Bank of Scotland) and with the clothing outlets served by Zoom.com.

It also seems certain to start an Internet war with Asda, the now Wal-Mart-controlled supermarket chain, which has prided itself on being among the first to move with Internet operations.

Tesco currently charges £5 to deliver goods ordered online but its managers are to become involved in 'online customer panels' to find out what people want from the service.

Reports in other parts of the world have suggested that one of the things they want is free, or reduced price delivery.

Carol Bradley, Tesco's E-commerce director, said the panels grew from a recognition that Internet shopping is a rapidly developing sector, and an area of 'constant learning'.

10 January 2000

Supermarket brands lag behind

A survey from Ernst and Young has shown that the major supermarkets are falling behind in the brand race on the Net.

Shoppers on the Net are looking for familiar brands but because Tesco (generally seen as the market leader in food retailing) and Sainsbury have not as yet organised national distribution, they are losing orders, according to Ernst and Young.

Brand awareness on the Net is becoming a critical factor in surfers choosing where to shop said Ted Yates, the retail and consumer spokesman. 'The main challenge today is not if, when and how to get online; it's now a question of how e-tailers can develop and augment their brand and position in the marketplace'.

Yates said supermarkets were low down on the list, due to many of their sites being at a regional trial stage. This was an area consumers, looking to spend less of their leisure time in the supermarket, wanted to see improved.

Products most purchased on the Net were books, CDs, computers and related products, with Amazon, QXL, Lastminute, Easyjet and Ticketmaster among the top sites for UK shoppers.

19 January 2000

Monster portal edges closer to launch

It's big, it's backed by TV, it's been planned for three years, now and it's almost here.

Efdex.com [http://www.efdex.co.uk] is threatening to swamp the food trade with one of the biggest and most advanced pieces of IBM technology yet seen.

Promising to be all things to all buyers and sellers, the glossy new site is claiming to be the future marketplace for the whole food industry. Details of who the all-day newscasts are aimed at and which sectors are coming online first are still vague. The business-to-business site promises market reports, prices, directories and instant secure purchasing. Offices have been opened in Reading, Singapore and Stamford, Connecticut.

So far we have only been able to admire the brochure and the ease of the graphics, but registrations are now being taken at enquiries@efdex.co.uk [mailto:enquiries@efdex.co.uk].

01 February 2000

Big two lashed in consumer test

Flushed with e-commerce headlines both Tesco and Sainsbury fell foul of a Daily Mail consumer test.

Forget the hype, when put to the test both Tesco and Sainsbury flopped dismally in consumer trials. The Daily Mail ran comparative shopping experiences on five major on line supermarkets.

- Tesco couldn't live up to its promise of next day delivery . . . and could only manage a five-day delivery turn-around in Oxfordshire.
- Sainsbury meanwhile took one hour and 40 minutes online to process a £26.67 order.

- Asda also posed a problem to Mac users, but redeemed itself by a phone ordering service.

Top of the survey came Iceland and the small trial site for Budgens.co.uk which only operates within 45 minutes of Harlow, Runcorn and Tamworth. Budgens even managed to cross the road to include a Tesco chicken which was the ordered size and topped up the basket with free chocolate.

16 January 2000

Attack on online supermarket prices

The Consumers Association has attacked the cost of buying online from supermarket brands.

Shops are not passing on savings to the customers: 'You would normally expect the Internet to offer cheaper goods and services,' said a spokesman, 'The feedback we have had about services is very mixed. Some people find the sites confusing.'

Tesco chief executive Terry Leahy who is currently said to be taking £2million a week online, refuted the criticism saying that Net shopping was about convenience not price.

16 February 2000

Sainsbury appoints Egg head

Rampant speculation is spreading through the Internet community that Sainsbury is about to unleash a whole new generation of online strategies.

Rumours have been fuelled by the appointment of Peter Davis as chief executive. Davis was head of the Prudential when it set up Egg.com, the online bank.

16 February 2000

Banks hushing up security attacks

Security specialist architecture company Mi2g.com is alleging that major financial institutions are not making public cyber attacks on their systems.

Following on from the widely publicised attacks on consumer sites (known in the trade as DDoS or Distributed Denial of Service), Mi2g [http://www.mi2g.com] alleges that more serious electronic attacks are launched on financial institutions than are being made public. The company – which is not without a vested interest, having originally been formed to set up car and lawyer portals (then known as lounges) before moving into security – is claiming that off-the-shelf software is not sufficient to protect companies from cyber attack.

It charts the fall in share price for victims of cyber attack at between 5–9% for the duration of the attack. It alleges some cyber attacks may cost the company as much as £1.5 million per hour in leached share capital.

Another theory on the current wave of cyber attacks is that the source may yet prove to be unscrupulous elements in the fourth estate. Journalists looking for a story may have paid friendly hackers to see how secure brands may be.

16 February 2000

Tesco is going to fall over say Americans

Flushed by its claimed success as the biggest online grocer in the world, Tesco has been blushing at American criticism of its technical system.

Aghast at the arrival of this new kid on the block, American web authors and e-commerce strategists are saying that the whole Tesco infrastructure is going to implode. This is an extract from the American based e-newspaper The Standard: 'The Tesco Direct service is so low-tech it's bizarre. Orders are sent to the Tesco store nearest the customer's home, then 'pickers' pull the desired items off the shelves. You read that right – no warehouse. 'When stores are crowded at weekends,' said the article, 'efficiency falls off further as pickers jostle their way down aisles and queue like any other shopper'.

Moreover, Tesco Direct runs on Microsoft (MSFT) BackOffice and cheap Dell

(DELL) servers, 'neither of which are really suited to the heavy-lifting of an e-commerce site supporting many thousands of concurrent users'.

17 February 2000

Tesco defends in-store picking

Tesco has finally gone public with its defence of in-store picking over a strategy for dedicated warehouses.

Russell Craig said: 'We chose a store-based picking system because Internet shopping for food demands a great leap of faith from customers. If they see the local store operating the system alongside normal shopping, it does not seem quite so alien. People are more likely to buy perishables if they come from a store where goods are turned over faster than a warehouse'.

Using the stores Tesco estimates that 95% of the country will be within a 25 minute drive. One picker can service six to seven orders at a time. Analysts, though, suggest it can cost £8–£12 per order picked in-store, against £5 an order from a warehouse, although American predictions from Webvan suggest this figure can come down even more with automated lines.

26 February 2000

Iceland warms to the Net

The 760-store chain is to follow up its commitment to nationwide delivery from Net orders with a £8 million re-branding exercise as Iceland.co.uk.

Chairman Malcolm Walker said: 'The new name says we are in the 21st century. It's our website [http://www.iceland.co.uk] address and everyone in Britain will know it. Even our most boring of advisers thinks it's a great idea'. Orders are currently running at 5,000 a week and Walker claims many of these are new customers who do not usually shop at Iceland.

In a further departure from Iceland's frozen image, Walker is to launch a full range of organic products in the autumn. He said the average online order was £62 compared to just £8 in the shop.

28 March 2000

M&S go all frilly

Marks & Spencer has put off its e-commerce development in food in favour of lingerie.

The crisis-hit market leader has taken a stake in Splendour.com [http://www.splendour.com], founded by Mark Pilkington, former chief executive of Gossard. David Norgrove, M&S Ventures chairman, said: 'What Splendour.com has is an Internet approach to retailing and driving down costs, which is something we can learn from'. Further investments in food, fabric technology and intellectual property are expected later in the year.

29 March 2000

Tesco to split off Internet arm?

Speculation grows that Tesco intends to split off its Internet trading arm as a separate company.

The success of the Net division of Tesco [http://www.tesco.co.uk], which is claiming to be the most successful grocer online in the world, has led to increased speculation that it is about to float the division as a separate company. It has been claiming to have around 10,000 customers signing up a week, with an average spend of £93, making it an attractive portal site for would-be investors. Other speculation has it that the supermarket giant would then be able to provide delivery for all online Net services and not just sell Tesco products.

07 April 2000

Sainsbury launches razzmatazz down the email

Sainsbury has launched a bid to win over new online customers with promises of prizes, competitions and reward points through their email.

The campaign which launched at the beginning of April promised 20,000 reward points for the lucky registrant in a lottery-style draw. There were also 100 bottles of wine to win and details of promotions, although the details

Efoodnews.com received were for offers that expired the same or next day. Also in the email came a recipe for thyme-encrusted lamb chops.

07 April 2000

Tesco's Net profit

The sales for the UK's largest supermarket recorded sales figures of more than £20 billion – £18 million at home, and £2 million abroad.

Tesco has also announced that it is separating its internet operations into a fully owned subsidiary. Pre-tax profits for the year rose 8.4% to £955 million.

Since launching its internet site [http://www.tescodirect.com/default.asp], the operation has rewarded the group with an annualised turnover of £125 million and is currently welcoming 200,000 visitors to its site.

Tesco has announced that it will be spending £35 million on the creation of a new unit called Tesco.com [http://www.tesco.com/].

The Tesco internet site was launched in 1996 and placed itself at the forefront of the 100-plus shops that have now moved online. By the end of the year, it is expected that 90% of major stores across the country will be operating online.

With home shopping set to reach 10% of the grocery market by 2005, the marketplace is wide open. Tesco's rivals are also stepping up the pace by establishing their own websites. Waitrose is aiming for 60,000 customers by the end of the year, while Sainsbury's has employed Sir Peter Davis, the brains behind the Egg online banking brand, to run their site.

14 April 2000

Dark horse headhunts logistics wizard

The Euro food start-up, LM Solutions, has fuelled speculation of a major intervention in the market by announcing new blue chip appointments.

The company, which to date has only said that it will 'focus on opportunities in the European food market' and has given itself more than a year to reveal its full plans, has recruited Robert Gorrie, formerly the IT director of trucking and transport group TDG.

LM Solutions was set up by two senior Marks & Spencer executives last month, both of whom were said to be dissatisfied by the retailer's approach to e-commerce. Nigel Robertson was Head of Change Management and Roger Whiteside was Director for Food.

Another founding member, Jason Gitting from the investment bank Goldman Sachs, said: 'We are a food retailer but we are also a logistics business: getting food to people's doors is going to be one of the most significant problems an Internet company faces.'

25 April 2000

Online grocers lashed in consumer test

Food and grocery deliveries have come out worst in tests on deliveries from supermarkets.

Two new studies, one by the Times and one by agency MCC International, both came out highly critical of the quality of online service. Graham Thatcher of MCC said that if anything service was 'getting worse' not better and highlighted food deliveries as bottom of the list.

The Times journalistic survey seemed to lack some Net savvy but was strongly critical of the length of time it takes to register. Worst for delivery was Marks & Spencer: three weeks to deliver a pair of knickers!

25 April 2000

Safeway snaps up home delivery company

The American giant which has been slow to move into e-commerce has bought a Dallas-based distribution company Groceryworks.com which may herald the dawn of a new strategy.

Safeway [http://www.safeway.com/] operates about 1,650 stores across America and has annual sales of about $31 billion. The deal involves $30 million in cash, plus goods in exchange for 50% of the company. The plan is to extend the service [http://www.groceryworks2.com/] to 16 metropolitan areas and build up a network of 21 fulfilment centres by the end of next year.

Analysts believe cybergrocers will grab a growing share of the $440 billion grocery industry as consumers place more value on convenience.

'If these forecasters are right, and it's 10% of the business by 2007, that doesn't seem like a lot but in our sector it's about $50 billion,' said Steve Burd, Safeway's chairman and chief executive. But sounding a note of caution Burd also pointed out that grocery and distribution are two low-margin businesses: 'This is a very, very challenging channel in which to make money,' he said.

25 April 2000

Now Tesco goes into mortgages

Feeding speculation that food grocers that set themselves up as portals can diversify into any areas they want, Tesco has unveiled plans to move into mortgages.

It is joining forces with the online arm of independent mortgage broker John Charcol to launch a mortgage supermarket on the Internet. Under the alliance with CharcolOnline [http://www.charcolonline.co.uk/], Tesco Personal Finance [http://www.tesco.co.uk/finance/home.asp], a joint venture between the supermarket chain and Royal Bank of Scotland, will offer a new Tesco Mortgage Finder service starting this month.

The service will provide market-wide mortgage search, selection and online application facilities, as well as Tesco 'special mortgages' – deals with British mortgage lenders only available through the Tesco Internet site [http://www.tesco.co.uk].

'The alliance marks the entry of Tesco into the UK mortgage market and forms a key part of its aggressive e-commerce strategy,' Tesco Personal Finance said in a statement which went on to explain that the new service would help its customers to save money and speed up the process of finding the right mortgage.

02 May 2000

Tesco looks to go overseas

Tesco is claiming to have signed up a total of 500,000 UK home shoppers – with an extra 100,00 added in April alone – and is now looking abroad to extend its reach.

Marketing director Tim Mason said that South Korea, Ireland and Taiwan would be obvious targets as Tesco was already building stores in those countries and Internet usage there was already high. He was also looking for joint venture partners in America. Last year Tesco.com had annualised sales of £125 million.

15 May 2000

Asda to launch TV and Web shopping at the end of the summer

Interview with David Martin of Asda@home.

The Asda [http://www.asda.co.uk/index.html] strategy to date has been a dual attack of a 98-page catalogue and a CD-rom, which has allowed the company to trial its depot delivery. Later this summer it will finally catch up with rivals with its own online service and TV shopping scheme. Surprisingly perhaps the focus is to be on food not general groceries and so far the take-over by American Giant Wal-mart has yet to influence its approach.

How do you see the future of the Net?

'We believe TV will revolutionise the sector.

What teething problems have you had so far in developing a mainstream Net service?

'There are huge issues of availability. When people order on a Monday the same goods are often not available on a Wednesday or a Thursday. But we think the jury is still out on the benefits of same-day delivery. Most customers say it is nice but not strictly necessary.

What are your key targets for the Net?

'We deliver in our own multi-temperature vans. Owning the last mile is a key part of our strategy. We always see ourselves as wanting to operate that last mile so that we can control our relationship with the customer on the doorstep.

What successes have you had as yet on the Net?

'70% of our Net customers have come from our competitors.

And what failures?

'Our understanding of our customers is very poor. We have no Club card like Tesco and so we have very little information. We see the Net as a way to allow us to learn more about our customers.

Why have you been so slow getting started?

'We spent £5 million on the Net last year developing our website and back end, which is about two thirds of what it would cost to open a new store. In the first week we took £65 where a new store may take hundreds of thousands of pounds. So it is simple economics really. Whatever our competitors say no one is really making money on the Net. Not yet anyway.

And what have you learned?

'When we first started in Croydon we recruited men because we thought we had a warehouse and needed warehousemen. But we have found that women are much more successful in this business. They are much more accurate in their picking and understand what the customer wants.

'We have also learned the importance of talking to people and developing our contact with the customers. So far it has been a fairly limited trial with a break-even of about 3500 orders a week.

Is your produce better if you pick it yourself?

'Considerably better when we sell it online. If you buy fruit and vegetables in the store then it may have been lying around for two days in the depot and then two days on the shelf. An apple may even have been handled 50 times. With dedicated stores hand picking for Net orders our produce is much better.

How does the Net spend compare to the store spend?

'On average a sale in store will be about £25. On the Net it may be £80.

What is your best selling line on the Net?

'The 24 pack of toilet rolls.'

02 June 2000

Whoops! Safeway takes down the photos on its website...

... because no one checked the copyright!

Safeway [http://www.safeway.co.uk/] has had to remove the photographs from its website on instructions from its lawyers. The digital copyright in the photographs did not belong to the supermarket. Angry photographers lined up with their invoices. 'It has cost us a packet,' said an anonymous spokesman.

02 June 2000

Jet and Londis sign up to new Dropzone concept

5000 stores are targeted to pick up email groceries.

A UK variation of New York's Kozmo [http://www.kozmo.com/] has been busy signing up petrol stations and corner shops as pick-up points for online groceries as the war for the last mile in the online wars starts to hot up. Dropzone1 [http://www.dropzone1.co.uk/] is the company.

02 June 2000

Sainsbury ties up deal with Carlton

Sainsbury and TV giant Carlton have signed a deal to develop online food and wine sales.

The marketing department has always led the way at the supermarket and the move will bring together one of the country's mightiest retailers with one of its most powerful media players who were an early convert to the benefits of the Net with SimplyFood. The web presence has been increasingly aligned with the Carlton food channel bringing the link between the Net and TV ever closer. Now with the Sainsbury tie-in the strategy looks like it may be reaching its logical conclusion. The broadcaster is also launching a £154 million pan-European Internet and interactive TV venture with counterparts Kirch, in Germany, Mediaset in Italy, TFI in France and Spain's Telecino.

02 June 2000

Milkman floats e-delivery

Express Dairies is looking to hook up their milk floats for Net deliveries and win the last mile of the Net race.

Express [http://www.expressdairies.co.uk/] has 150 depots and operates a fleet of 2500 floats with a further 1500 run by franchisees. They are cheap to run and environmentally friendly. 'We already deliver to 2.2 million customers every day but we can handle a lot more than that,' said chief executive Neil Davidson. A tie-in with an electrical company is expected soon to deliver hi-fis and TVs.

More interesting is that negotiations are going on with DHL and Federal Express to handle 'the last mile' of the e-economy.

The announcement coincided with a fall in traditional profits from £60 million to £52 million.

07 June 2000

The rise and fries of company shopping

Interview with Bill Pratt of Waitrose@work.

Bill Pratt is responsible for rolling out the innovative and successful Waitrose@work business. He rose up through the corporate ranks from an area manager. The programme converts companies with more than 300 desktop terminals and allows employees to order online and get their groceries to take home.

How did Waitrose@work get started?

The idea of having one van that could serve 80 homes seemed appealing. The first pilot was launched at Canary Wharf in 1998. It was a good vehicle to promote both Waitrose and Canary Wharf, and it allowed us to stay ahead of the competition. We began with just 500 lines, then launched to a full 10,000. British Airways was our first customer. Then Microsoft came along with three sites. Last year we had 30. Currently we have 44 sites. We are now putting on about four or five a month. We could do more, but we feel it is better to grow slowly and steadily.

Who uses the service?

A very high percentage of shoppers have never shopped at Waitrose before, so the Web has been very effective in converting new customers. And because we pick in-store, it is a good use of bricks and mortar too.

Do you envisage going over to specialist depots?

With the potential we have in existing branches I cannot see that we want to get into dedicated warehouses.

How do companies respond to the idea?

It is a very good way to build good relationships with our corporate customers. It breeds loyalty. I am surprised none of our competitors are doing this. Some of our customers now include Hampshire County Council, Goldman Sachs and the London Business School, so there is no stereotype company. It covers the spectrum. Currently we have 25 branches trained up to service companies and each one can handle about six companies. We do a lot of promotions on-site like wine tastings. To be successful we need to get as many people to use it as possible.

And employees?

About 50% of the workforce usually registers. And then about 20% will use us weekly. Some of our smaller sites have been very successful just because of the number of people in the company who use us. For example John Brown Publishing, who do our magazine, is a very successful site, where the BBC at White City is huge but not necessarily so effective.

What differences are there between sites?

The transactional rate in London is much lower than in a greenfield site. Outside of London a sale could be £40–£45 an order but in London it may be half that. It is a stress-beater.

Are you making money?

We will be.

08 June 2000

Safeway trial Palm 3 software

Safeway is trailing new software developed by IBM to convert Palm 3 pilots for handheld shopping.

Jeremy Wyman from the IT division has revealed that the supermarket is testing a revolutionary new approach to shopping. 'Most customers only ever buy between 120 and 150 items. The Palm can carry those easily. Plus it will be able to remember orders for the previous five months. It takes about three minutes to train someone how to use it. Soon it will be available to download directly from the Web.

'Most of our customers prepare a shopping list through the week. And then they forget things. They usually know how much they want to spend, how many calories they want and how much they want to carry home – but invariably they overlook something. With the Palm pilot they will have an easy reminder system'.

Trials have also been going on to check undercover marketing. Data on like-minded people is assembled and reminders are sent out through the email to suggest new things customers might like to try. 'It seems to work,' said Wyman.

08 June 2000

LM solutions snaps up tracking software

Further clues to the shape of the e-commerce company set up by former Marks & Spencer executives have emerged with news of a top line deal with American Nasdaq listed software house Descartes.

LM Solutions [http://www.lmsolutions.com/home.htm] nipped ahead of both Sainsbury [http://www.sainsbury.co.uk/] and Tesco [http://www.tesco.co.uk/] in securing a deal for Descartes [http://www.descartes.co.uk/] delivery Net software. This allows for tracking of goods for home delivery as well as pin-pointing timetables. It is used in America by Webvan, Ericsson and Coca Cola [http://www.cocacola.com/alternate.html] among others.

Speculation is mounting that the company will not just launch an e-supermarket next year but may create a delivery network for other retailers to use at the same time.

20 June 2000

Express buys into M-box

Express Dairies has expanded its e-commerce ambitions by buying into a fledgling home shopping and fulfilment company M-box.

Last week Express [http://www.express-dairies.co.uk/] said it planned to use its fleet of 4000 milk floats and 150 depots to develop delivery systems for e-commerce. The initial targets will be London, Birmingham and Manchester, but in total Express says it could deliver to up to seven million homes. Director Tim Smith said: 'It's old economy meets new economy'.

20 June 2000

Why Somerfield had to drop its e-commerce

Vaunted as one of the pacesetters last year, Somerfield has had to drop out of the e-commerce race.

Plans for Somerfield [http://www.somerfield.co.uk/index.asp] shoppers to order online or by fax or phone have been dropped, although home delivery will continue, chief executive John von Spreckelsen has announced. Analysts said the move would damage Somerfield's future sales growth. Mike Godliman of retail analyst Verdict Research [http://www.verdict.co.uk/] predicted 5% of supermarket food sales – £4.96bn – would be done online by 2005. 'That's a big chunk of the market to ignore.'

Somerfield's 24-7 has been accumulating losses since it began operating last September and last month estimated annual turnover at £4m. The company did not deny speculation that two of its three 24-7 fulfilment centres received as few as 15 orders a week.

'We may well have a fresh look at the role of e-commerce food retailing within Somerfield in the future. But the hard fact is that 24-7 has not been growing and represents a significant distraction to management at a time when its focus has to be on the core Somerfield and Kwik Save business,' said Von Sprekelsen.

20 June 2000

The Web's cup runneth over

Now even BT wants to sell wine.

Not content with its phone monopoly BT has announced plans to launch a wine service with ItsWine.com [http://www.itswine.com/], which is claiming to be one of the oldest wine websites on the Net. The company has been behind wine shops for companies as various as Argos, Iceland and even Yorkshire Cricket Club.

BT plans a dual service, one being tied into its new faster ADSL line service which is launched in London. Prices at the bottom end start at £2.99, with delivery another £2.95. Director Edward Mills claimed new technology would allow it to provide the 'most sophisticated, interesting and robust service' on the Net. Where have we heard that before?

04 July 2000

Coffee merchant snaps up mall

In a surprise tactical move, a coffee brand has swooped in a shares deal to acquire one of the best known malls for small producers.

The high-street brand Whittard has paid £1.2 million to buy Bestofbritish.com [http://www.Bestofbritish.com], a mall that includes other well known high-street brands such as Bendicks and luxury goods retailer Mulberry. The mall was owned by Holinger Telegraph New Media, the online investment group owned by the Daily Telegraph.

Whittard has been seeking brand associations with other specialist food names for the last few months in a bid to develop its portfolio and spread its online sales. BestofBritish.com is one of the longer standing malls having originally established a trading platform with sales to expatriates.

10 July 2000

Part 6
USA Models

America is the home of the Net and as quickly as the UK and the rest of the world, especially Japan, catches up, the first generation of major portals and web initiatives came from across the Atlantic.

Credit card fraud down to 2%

A new survey in America for Shop.org shows that only 2% of customers have experienced any online thefts.

However, the survey has underlined also that 75% of people questioned insisted that security was the single major issue in buying off the Net.

It is not clear from the figures whether the 2% have actually been involved in a fraud, have had disputes with companies, or if the fraud was someone else in the family using the number. More disturbing is that research from International Data Corporation says that one in five large companies (those having 1000 or more employees) is likely not to have a secure transaction available. In small companies that ratio rises to one in three. Hardly likely to instill customer confidence.

04 January 2000

UK catches up on USA

The UK e-market is now second only to that of America according to new research with online spending tripling in 1999.

Similar, but not so spectacular rises are recorded in France and Italy, Ernst & Young said in a report. Already 10% of U.K. households shopped online in

1999 and British Internet spending was set to treble in the next two years, the report said. 'Our report shows that online selling is a real and viable strategy for growing total business revenue,' said Ted Yates, Ernst & Young's retail and consumer practice spokesman.

Significantly he said the report indicated that online sales did not eat into sales from other outlets, instead they built the top line. 'For a merchant, going online is not just a market protection strategy – it's a growth strategy,' he said.

Competitive prices, convenience and a wide range of available items were the main reasons driving Internet shopping, the report found. But consumers pointed to issues retailers should address such as total credit card security, guaranteed privacy for personal information gained from online transactions and a need for wider choice.

10 January 2000

The tills are ringing in USA

Santa Claus came a week early for e-traders in America this Christmas.

First figures out of America show that there was a 270% increase in trade in the week before Christmas over last year. A staggering 27% extra visitors checked in to e-shopping sites in the critical week before the holiday. AOL was claiming its customers bought $2.5 billion worth of goods online, almost a quarter of the total annual turnover. And online customers were tending to spend freely – usually an average of six purchases per credit card, at an average value of $134 per spend.

But not everyone was ready for the surge. Ernst & Young said its year end survey of 1,283 Internet users found that Amazon.com was given the best report card by consumers while Toysrus.com once again 'got a Yuletide raspberry for failing to deliver the goods'.

Meanwhile, online seller of computers and office equipment Online America announced lay-offs and a sale of the company, having over-extended itself.

19 January 2000

Food lags behind in USA

Sluggish figures for the pick-up of efood businesses have come out of America.

Research figures have shown that in 1998 only 1%, or $235 million, of the $720 billion grocery market has moved across to the Net.

The most successful company, Peapod [http://www.peapod.com/], has taken ten years to amass a base of 100,000 customers and it admits to still losing money in five out of eight of its major markets.

Poor service, high prices and bad selection are the reasons most customers said they did not want to use the Net for general shopping. The USA model is compounded by the logistics of delivery across much wider areas than in the UK, where the market conditions have been shown to be different.

07 February 2000

Net shoppers demonstrate brand loyalty

A new study has shown that Net shoppers stay with brands provided they get good service.

An American survey of 200 shoppers showed that more than half were happy to stay with familiar sites and changing was regarded as a nuisance. Most cited how good the shopping experience was on the Net as a key decider. The information from Cognitative Inc put fast response time at the top of the list with 37%, but only one point behind was familiarity.

Things likely to upset web shoppers were outdated information and poor customer service.

07 February 2000

Webvan gets $120 million backing

A scheme for a new Net-based Internet supermarket with home delivery has received huge backing from funders.

The Webvan [http://www16.webvan.com/] scheme aims to create the last mile of e-commerce by delivering to the door, according to founder Louis Borders. He has just raised $122 million to create an infrastructure that will allow high quality fresh foods to be delivered within a 30 minute window.

More than 10,000 households have signed up to the Webvan service since it launched in the San Francisco Bay area in June. Webvan has hired some 80 software programmers to create proprietary systems that automate and link every aspect of its business processes.

'This is the first backend re-engineering of an entire industry,' says Kevin Czinger, Webvan's chief financial officer, a former managing director at Goldman, Sachs & Co. Gone are the scores of stock personnel, warehouse employees, and cashiers that grocery stores typically employ.

At the heart of all this re-engineering is the 330,000-square-foot distribution centre in Oakland, California, which services an area of 40 square miles in any direction. This is the prototype for the other centres that Webvan plans to build.

The $25 million facility, which includes 4 1/2 miles of conveyor belts and temperature-sensitive rooms to store items such as wine, cigars, and ice cream, can serve as many customers as 20 supermarkets, which typically measure about 40,000 square feet. The difference is that Webvan can do all this with less than half the labour and nearly double the selection of available products. Webvan eventually plans to sell about 50,000 items, including 300 varieties of fresh fruit, 750 kinds of cheese, 500 types of cereal, and 700 cuts of fresh meat and fish, including live lobsters. A typical grocery store carries an average of 30,000 items.

Twenty-six more markets at a cost of $1 billion are scheduled for the next two years. Financially Webvan believes it can reduce the overhead cost to 1% from a traditional 6% and increase margins by 10%. Webvan claims to be able to assemble an average 25-item order in less than one hour.

At peak performance, the company expects to bring in $300 million in annual revenue per facility by handling more than 8,000 orders a day, involving a total of 225,000 items. The average conventional supermarket brings in about $12 million in annual revenue.

High-tech tracking systems monitor orders from the moment they are placed on the company's Website. Workers known as *pickers* are stationed throughout the distribution centre to assemble orders in plastic boxes known as *totes*, which are colour-coded depending on whether items are refrigerated, frozen, or dry. But instead of traipsing down endless aisles, the pickers travel no more than $19\frac{1}{2}$ feet in any one direction to reach 8,000 bins of goods that are brought to the picker on rotating carousels.

'When 85% to 90% of a person's time [at a traditional store] is spent travelling to locate and assemble orders, you realise why this makes sense,' says Gary Dahl, formerly a senior executive at American Stores Co. and now Webvan's vice-president for wholesale. 'You want to keep people stationary so you can keep them busy'.

Once a picker has finished with the task, the tote is then transported via conveyor belt to other areas of the facility housing different items. After an order has made the rounds, it is loaded onto trucks refrigerated to 35°F (c. 2°C) that take it to one of 12 docking stations throughout the Bay Area. From there, the totes are loaded onto one of the company's 60+ vans so that drivers can take the orders directly to the customers' homes. None of some 70 'couriers' travels more than 10 miles away in any direction.

08 February 2000

Coffee is the froth on top for e-newsagent

Starbucks coffee houses have woken up $150 million richer thanks to an e-commerce tie in with an e-newsagent.

Kozmo.com [http://www.kozmo.com] is a city-based home delivery service that specialises in traditional newsagent-style products from videos to playstations, magazines, snacks and newspapers. It has announced a $150 million tie up with the Starbucks coffee chain and hopes to be delivering hot coffee to offices and homes before the end of the year.

The alliance puts the Kozmo brand name in front of ten million Starbucks consumers, many of whom shop weekly at one of the chain's 2,400 stores. 'Riding on the coattails of the Starbucks brand will be a tremendous benefit for us,' Kozmo founder Jo Park says.

The deal will bring more than 1.5 million new customers to Kozmo. 'That was key to the alliance', said Park. Customer acquisition costs for many e-commerce startups have been hovering around the $100 range. Kozmo, says Park, will be 'linked with one of the most trusted brands' as it enters 15 new markets by the end of the year. Over the five-year span of this deal, Park estimates his customer-acquisition costs could drop to as little as $50 per customer.

For Starbucks [http://www.starbucks.com/], the deal helps fill a glaring hole in the company's very public e-commerce aspirations. The company has been selling an assortment of items online, from coffee beans to New Age CD compilations. It has also aligned itself with other e-commerce vendors, such as Cooking.com [http://www.cooking.com/], to sell small appliances such as coffee grinders. But the company had no recognisable strategy to sell its big moneymaker – those $2 coffees and $3 grande lattes. 'This alliance defines the benefits of a truly integrated *clicks-and-mortar* strategy for our customers,' said Howard Schultz, Starbucks' chairman and CEO. 'In fact, the similarities between Starbucks and Kozmo customers make this a value-enhancing proposition for everybody.'

According to a press release, 90% of the Starbucks' customer base use the Internet. Kozmo will become the exclusive Web delivery service for Starbucks. In the beginning, Starbucks will use Kozmo's bike messengers to ship non-perishable items like coffee beans to customers' doors in under an hour. Later this year, Park says, the companies will test the delivery of hot coffee. In addition, some Starbucks stores will host Kozmo drop boxes, enabling customers to return movie rentals more easily.

Kozmo currently operates in five cities: New York, Boston, Washington, San Francisco and the Starbucks hometurf of Seattle. Park says Los Angeles will soon become the sixth market.

17 February 2000

Grocer pays $60 million for AOL link up

The home-delivery online shop HomeGrocer.com delivers in four USA cities at present, including Los Angeles.

It will use the promotion to develop into another ten cities including New York. Under the five-year agreement, AOL users in HomeGrocer's [http://www.homegrocer.com] delivery areas will be able to reach the grocery service through the AOL.com, Netscape Netcenter, Digital City, AOL MovieFone and Oxygen.com websites.

The company said last month that a planned initial public offering could raise as much as $264 million, $14 million more than originally projected. The move in the UK will be seen as justification for Tesco's Net strategy of becoming a portal site in its own right.

22 February 2000

Chocolate site aims for $5 million

The man who set up Cobra beer has moved into online chocolates.

Planetchocolate.com [http://www.planetchocolate.com] will be the first network of chocolate retailers in America with plans for global expansion by the end of the year. Entrepreneur Arjun Reddy, who previously launched the popular Cobra Beer in the UK, aims to sell $5 million in the first year and grab a slice of the $23 billion Americans spend on chocolate and confectionary alone.

Planetchocolate.com already has over 100 chocolatiers on its network, spanning Boston to Hawaii, and will act as a one-stop-shop for virtually any chocolate and confectionary product.

Inquiries have come from chocolatiers as far as Paris, London and Dublin: 'A number of retailers have their own websites but people quite often don't know they exist. By joining up with us, they are bound to get business traffic,' says Reddy.

Stores can sell up to 50 different items on their webpages hosted by planetchocolate.com, from straightforward candy bars to specially gift-wrapped boxes of your favourite confection.

09 March 2000

Tavalo secures $36 Million in funding

The online retailer and destination for consumers who cook, dine and entertain, has announced that it has secured an extra $36 million in venture capital and institutional investor financing.

Van Wagoner Capital is funding the investment. Garrett van Wagoner said: 'In today's volatile market, you look for companies that have a strong business model, a large addressable market and attractive margins, multi-channel operations, strategic partnerships and a strong management team. Tavalo [http://www.tavalo.com/] has clearly demonstrated it has all the right ingredients and is well positioned to become one of the standout companies within the lifestyle retailing category.'

The company has also announced a three-year agreement with American Express Publishing. Mark Sanich, Senior Vice President of Marketing for American Express Publishing, said: 'This alliance provides wonderful opportunities to offer unique promotions and valued content geared to the millions of consumers who share a passion for all things epicuran. We're also pleased to be able to offer extraordinary benefits from Tavalo's rich content and product selection to the exclusive members in our Food & Wine Connoisseur Club.'

Kevin Appelbaum, CEO for Tavalo, said: 'These new funds will allow us to continue enhancing the products and services we provide to Tavalo customers. Our new strategic alliance with a leader in the culinary lifestyle market will help further differentiate Tavalo in a rapidly growing and changing market place.'

14 April 2000

Online meat exchange draws record response

American website FoodUSA.com has launched an online meat exchange.

The site [http://www.foodusa.com/] has reported that 518 orders have already been placed as the streamlining of the buying and selling of wholesale meat products begins.

The exchange uses a simple bid/ask system to bring buyers and sellers together in a neutral exchange environment. In its first day in operation, a total of 17.6 million pounds of meat and poultry valued at $19.2 million were offered.

Tom O'Connell, FoodUSA.com Chief Marketing Officer, said: 'This result demonstrates an eagerness within the meat industry to capitalise on the efficiencies and reduced costs offered by FoodUSA.com's e-commerce solutions.

'We are extremely pleased by our launch results and anticipate rapid growth as more and more meat companies learn of this success.'

FoodUSA.com has received the endorsement of the American Meat Institute, which has recently joined the company as a partner dedicated to developing e-commerce solutions which will increase the efficiency, profitability and safety of meat and poultry trade worldwide.

14 April 2000

Cybercourier wars set to hit London

Three new door-to-door delivery ventures look set to go head-to-head in London.

The big players are American companies Kozmo and Urbanfetch both of whom have trialed systems in New York. Going up against them will be the home grown service Queuejumper.com [http://www.queuejumper.com] which promises 'mouse to house in half an hour'.

Kozmo [http://www.kozmo.com/] is raising $100 million on Nasdaq to extend its operations into Europe. Currently it operates in New York, Seattle Boston, San Francisco and Washington. Recently it announced a marketing deal with Starbucks coffee chain.

Urbanfetch [http://www.urbanfetch.com/p/f/main.asp] only launched in October last year in New York and is claiming 1000 deliveries a day in Manhattan. It has appointed advertising agency St Luke's to spend £3 million this year and a further £7 million next to win market share. The bulk of home deliveries are CDs, returned videos and snacks and groceries.

Thomas Fitch of the London-based Queuejumper.com said: 'The service relies on people's laziness. We charge 90p for delivery and collection and people will generally spend that rather than go to the bother of collecting things themselves.'

25 April 2000

Big retail shakeout predicted

Respected American research agency Forrester is forecasting a major shakeout of online retailers.

Many first generation sites are 'dead shopfronts walking', says the report which then cites the tribulations of US players like CDnow.

Senior analyst Joe Sawyer said: 'It's time to face facts: Online retail's honeymoon is over'. He argues the problems faced by companies like CDnow and grocery chain Peapod are to become widespread: 'Financial turbulence and new competition will dry up venture funding and accelerate the dot.com shakeout as the year progresses.'

He predicts that consolidation among companies selling commodities that are popular amongst Web consumers (books, software, music) will occur by the autumn, and the big winners will be bricks and mortar businesses looking to snap up ailing e-ventures.

Merchants selling high-value, branded products such as furniture and clothing can expect stability, but only until 2002. 'The online retail model is all about volume. In the US market there are literally hundreds of near identical sites selling things at the same prices,' said Forrester.

Europe, in contrast, has fewer competitors for online retail, but has problems of its own: one being a very fragmented market. Pan-European retailers are set to dominate as national retailers effectively put a cap on their own growth.

The report argues that even the largest market (Germany) will only account for 26% of online retail sales by 2005. Only those firms that adopt a pan-European approach can expect to raise the kind of VC funding necessary to build a brand in the maturing market.

08 May 2000

Farm wars take shape in US

The fight to win a share of the projected $1 trillion agricultural seed and machinery market is unfolding.

Goldman Sachs estimate the market will explode before 2004. And in what is looking like an increasingly typical showdown the players are split between dot.com start ups in one corner and the might of conventional agricultural companies forming new e-commerce alliances in the opposing corner.

Farmbid.com [http://www.farmbid.com/] has already registered 90,000 customers to use its site to buy and sell seed, chemicals, machinery and other agricultural material. Other sites in the US trying to secure a slice of the farm pie are Directag.com [http://www.directag.com], Xsag.com [http://www.xsag.com], SellMeat.com [http://www.sellmeat.com], and Farms.com [http://www.farms.com].

But major agriculture companies such as Cenex Harvest States Cooperative, Cargill and DuPont are also moving in on the newcomers with an exchange due to launch this month called Rooster.com [http://www.rooster.com]. The beleaguered genetically modified food pioneer Monsanto has also set up Farmsource.com [http://www.farmsource.com] as a farmers' newspaper online, which may also prove to have been a tactically shrewd move.

15 May 2000

American grocery market set for hypergrowth

A new study suggests the online grocery market in America is about to rocket from $200 million last year to a staggering $8.8 billion by 2004.

The IDC, a research company based in Framingham MA, also predicts that the cost of market entry will be more than $25 million to create appropriate fulfilment centres. The bad news, according to Jim Williamson of IDC, is that the new generation of Net grocers will still face the traditional headaches of business. 'The new market shares many issues that have troubled offline businesses for years including capital efficiency, obstacles to market penetration, and the price of gas', he said, 'To everyone who gets involved it won't be an easy process, and the road to a profitable online grocery market will be littered with bankruptcies and broken hearts'. Other research has indicated that 43% of consumers are more likely to shop with their regular supermarket than not.

The UK market is currently predicted to be about two years behind the USA notably held up by the lower conversion of women as PC users, although the Tesco experience and also that of Waitrose@work indicates that possibly the UK market may not follow the USA blueprint as precisely as in other areas.

30 May 2000

Handy maps for the delivery man

An American software company has moved into Europe with its road map service designed for home delivery.

The Roadnet.com [http://www.roadnet.com/] service tracks the quickest and easiest route to any home and also allows for management tools to collate orders, create holiday rotas and meet time windows.

07 June 2000

Health police nail dirty restaurants on the N

In a radical move the New York health inspectors have set up a website to name names and identify dirty kitchens ... including some of the city's best known restaurants.

Some of Manhattan's most famous names have fallen foul of the New York City's Department of Health and had their dirty linen paraded at http://207.96.244/webfood/index.htm [http://207.96.244/webfood/index.htm].

Among the well known names are Nobu for using 'bare hands' in the preparation of sushi while both Chanterelle and Balthazar are shamed for having eggs on the premises. But contract caterers Somehow and Aramak have also been given unwanted bouts of publicity over temperature controls.

07 June 2000

Ranching portal looks to customise the news

A content-led B2B portal for farmers has developed software to localise the news and advertising.

Editorial Director of AgWeb.com [http://www.AgWeb.com], Bob Coffman said: 'We will know up to 35 fields of information about any given AgWeb.com registered user who is professionally farming or ranching. By knowing where he lives, we can deliver customised and localised weather conditions and forecasts; we can deliver his local cash bids for produce, and we can narrow both market coverage and news coverage to commodities he has told us are most important to his business.

'Each user's site will be built dynamically 'on the fly' to reflect his business interests. And, additionally, the user will see commodity-specific news and analysis through assignment to one of several Special Interest Groups, which will include discussion areas for that particular topic.

AgWeb.com is primarily a content-driven site, but with an e-commerce component of classified capabilities and storefront opportunities. Coffman said: 'We believe the jury may be out right now on full-fledged Internet commerce, especially in agriculture. We will continue to research and measure the trends, and will retain an e-commerce strategy in our business growth plan.

'We believe narrowly focused niche B2B arrangements make sense and hold promise. But such activity can burn up cash quickly in the process of figuring out what works and what does not. Our initial efforts will remain with aggregation of user demographics and targeted delivery of content and commercial messaging, through a site accessed multiple times each day by users needing real-time business information.

'In background, we should also be able to track which news items appear to be of most interest to the user and which are receiving the highest readership. This will help us detect 'content fatigue' on certain topics, or direct our editors to increase coverage on certain high-interest topics.

'If done right, the 'gee whiz' will be totally transparent to the user as the site subtlety delivers things which seem to be of highest interest to any given user. He may think this is the best site he has ever seen for dairy coverage while another is impressed with its coverage of the cotton industry, for example. No longer will website users have to accept a Mulligan's stew of articles in the hope that there is some meat there for them.'

14 August 2000

Priceline moves on Europe

Name your price pioneer Priceline.com is set for a European launch this autumn with air fares top of the list, but grocery waits in the wings.

The two-year-old company set up as a budget travel service but now sells anything from breakfast cereals to financial services. Chief executive is Timothy Brier the former m.d. for TWA for the UK and Ireland. He said: 'Priceline is all about price. At the core of it, we're a marketing company that uses technology (like the Internet), our intellectual property, and our patent on the buyer-driven model. In the US we're doing an average 100,000 flight tickets a week and around 40,000 hotel rooms.

'We've built a price-management tool for our suppliers. The seller can feel more comfortable about using price to shift their product through us. We are a brand-shielding mechanism for the brand neutral consumer. To start with, we are covering air travel, hotels, car rental and telephone minutes'.

Income has topped $352 million. Growth in repeat customers and reduced advertising and marketing costs helped to more than triple revenues from the same period last year. 'We believe we are rounding the final turn and [are] on the homestretch toward profitability,' Priceline President and Chief Executive Officer Daniel H. Schulman said. Schulman noted that the company added 1.5 million new customers in the quarter, bringing its total customer base to 6.8 million. Schulman said, however, that the 39 percent customer repeat rate, up from 26 percent a year ago, was even more important in improving results.

The company's revenues of $352.1 million during the quarter are 216 percent better than revenues of $111.6 million during the same quarter in 1999, and the $4.4 million operating loss was an improvement from a loss of $15.8 during the year-ago quarter. Net loss for the quarter amounted to $1.6 million, versus $13.9 million in 1999's second quarter.

Priceline.com [http://www.Priceline.com] is one of the targets of Orbitz.com [http://www.Orbitz.com] , a controversial consortium of 30 airlines that aims to offer discount airline travel tickets online.

14 August 2000

American pilot B2B audit system

A new study aims to validate the real time claims of B2B transactions.

The research will demonstrate the business value of aligning product and pricing data between foodservice operators and their distributors and between distributors and their suppliers. The pilot will involve Brinker International's Macaroni Grill, Levy Restaurants, B.E. Keith, Kraft Foods, Inc., General Mills and Tyson Foods, Inc.

'In order to take full advantage of today's increasingly sophisticated B2B capabilities, as a starting point trading partners need to communicate up-to-date product and pricing data in real time and in electronic formats their trading partners can use,' said Mark Allen, executive director of the project called Efficient Foodservice Response. 'This study will demonstrate the positive results that can be achieved by exchanging accurate, timely item and price information.'

'For operators, especially independents, this will be a real breakthrough,' said Andrew Van Ermen, director, information services for Levy Restaurants, who serves on the EFR Electronic Commerce Committee. 'This pilot will demonstrate that technological barriers inherent in the current business process can be overcome, thereby putting even the smallest operators on an equal footing with much bigger players.' The expected benefits for study participants, and for companies that embrace item and product synchronization in the future, will include:

- Reduced purchase price variances
- Fewer shipment mistakes due to item number errors
- Reduced invoice errors
- Less time to process invoice errors and credit memos
- Shorter invoice-processing cycle time
- Reduced processing time
- Increased product data quality

- Fewer order entry errors, and less time required to correct them
- Knowledge of successful e-commerce practices

For more information: www.efr-central.com [http://www.efr-central.com].

05 September 2000

Foodbuy.com consolidates USA food service domination

The $17 million funded site snaps up rival Foodservice.com and plans to launch open trading marketplace.

'Foodbuy.com [http://www.Foodbuy.com] has become the first and only portal offering a total foodservice business solution: content, community, group purchasing and data management services' said Scott MacLellan, President and CEO. 'As the single largest e-service opportunity in the $150 billion US foodservice market, we differentiate our services from competitors by streamlining the industry's fragmented and inefficient purchasing processes to deliver dramatic cost-savings.'

The Atlanta-based company claims more than $1.4 billion in aggregated purchasing power.

12 September 2000

---

200 jobs go at Food.com

Drastic re-think on pioneering USA home delivery.

One of the pioneering home delivery services online in America, which then diversified into listings, has sacked half of its 200-strong staff and is to close its Seattle office. Operations and financial directors John Laing and Joan Varrone have resigned.

Chief executive Rich Frank says he is intending to focus on the more traditional restaurant side of the business and recruit more chains.

21 September 2000

Gourmet sites merge

Venture money moves on the gourmet market.

GourmetMarket.com [http://www.GourmetMarket.com] has merged with VirtualGourmet.com [http://www.VirtualGourmet.com] with money from investment group iiGroup.

Neil Swartz, Chief Executive Officer of iiGroup, Inc. said: 'At the beginning of the year, iiGroup made an investment into GourmetMarket.com and took over the day-to-day operations of the company. GourmetMarket.com has a great foundation with over 150 co-marketing deals in place, and averages 320,000 unique visitors a month. We are very impressed with RCC's background and the companies they have built in the past, and we believe that combining GourmetMarket.com's strengths with those of VirtualGourmet.com will create a company that is a profitable market leader.'

Derek J. Rundell, co-founder and co-chief executive officer said: 'The combined strength of VirtualGourmet.com and GourmetMarket.com will allow us to provide the most extensive gourmet food offering throughout the online marketplace. The existing customer base, rich content, unique customer acquisition strategy, and new distribution initiative will propel GourmetMarket.com into a market leader position within the online gourmet and specialty food niche.'

GourmetMarket.com is a leading technology and e-commerce company that sells premium, hard-to-find food, wine and kitchenware products.

18 October 2000

Watching the harvest online

Wine site sets up web cams in vineyard.

WineToday.com [http://www.WineToday.com], an online publication from New York Times Digital, has launched HarvestWatch 2000. Surfers can log onto the site and travel along with experts to the top international wine estates for an inside 'peak' into the harvests – and personalities – of each. This year's coverage will span the globe from the European wine regions of Bordeaux, Burgundy, Tuscany and Piedmont, to those in Northern California and Oregon, providing exclusive details and insight into how the first vintages of the new millennium are taking shape.

Reports are supported by commentary from New York Times wine columnist Frank Prial and WineToday.com editor Tim Fish. Features include live web cam images broadcast via HarvestCam(TM), and 360-degree panorama tours of famous wineries from the properties and cellars, such as Chateau Haut Brion in Bordeaux, and Marchesi de' Frescobaldi's Castello di Nipozzano in Tuscany.

18 October 2000

Chicago raids the European larder

American gourmet site aims to grab European foods and sell them to America.

4appetite.com [http://www.4appetite.com] claims to appeal to everyone from novices in the kitchen to the most passionate European gourmet. It claims to have searched through Europe's best stores, kitchens, restaurants, markets and bakeries to bring back hard-to-find ingredients such as: Belgium candies, French truffles and cooking wines, Italian sauces and olive oil, Norwegian salmon, Spanish olives and peppers.

Each page also includes the 'Recipe of the Day' which is authenticated by the 4appetite.com team of European buyers from each country. The site is the brainchild of a mysteriously named Ms. Sak, of Chicago, a graduate of Roosevelt University with a degree in International Business. She has lived and travelled, allegedly, extensively throughout Europe before moving to the US. This is her third start-up and 4appetite.com will be her second successful website.

20 October 2000

More retail casualties as the American market gets tough

One of America's oldest online grocers Streamline.com has shut up shop despite deals with big names like Kraft and Sara Lee.

Chief executive Tim DeMello of the Massachusetts-based company blamed investor apathy and weak consumer demand. It follows other reports of underperforming trading sites. In the same week Shoplink.com closed and Peapod and NetGrocer carry on thanks to European funding. Also Webvan followed up its acquisition of HomeGrocer with the announcement of its own streamlining – of staff. The industry view is increasingly that the liquidators are going to be busy in the coming year as the web churn sorts out the aspirations from the achievers.

17 November 2000

Named and shamed

Seattle and neighbouring King County restaurants will now have their safety violations published on the Net for all to see.

'Our purpose in posting restaurant closures online is to keep the public informed about food safety problems at local restaurants," said Dr Alonzo Plough, director of Public Health for Seattle & King County.

The first restaurant listed is Chili's at 15600 N.E. Eighth St in Bellevue. It was closed from 3 p.m. Oct. 26 to 10:40 a.m. Oct. 27 for failure to correct repeated violations, according to the website. Workers failed to use proper utensils to minimize bare-hand contact with ready-to-eat food.

Public Health typically closes about 100 food service operations each year until their food safety problems are corrected.

17 November 2000

---

e-vineyard slips through the wine ban embargo

As President Bill Clinton ratifies legislation to allow states to ban shipments of wine, America's leading wine site evineyard.com announces it can carry on trading.

Statutes in every state require licences at every level: wineries, wholesalers and retailers are all licensed. 'eVineyard is, in fact, a licensed retailer,' said Larry Gerhard, CEO of eVineyard. 'Most of our would-be competitors are unlicensed. And any company selling wine without a licence is placing itself above the law,' he said.

'Our competitors say they are not in the retail wine business,' said Gerhard. 'They are operating as 'buyer's agents.' As such, they solicit orders, take orders and cause wine to be shipped to fulfill those orders. By legal definition in every state statute, these companies are retailers, regardless of what they call themselves. And as retailers of wine, they must be licensed to operate within the law,' he said.

'eVineyard's unique business model has given us a valuable competitive edge as the nation's legislative bodies take steps to enforce liquor laws that have been violated by a vast number of would-be players in the space,' Gerhard said. 'While competitors struggle to bring their business models into full legal compliance, eVineyard will continue to give its customers the opportunity to buy wine and wine accessories online.'

eVineyard [http://www.eVineyard.com] legally serves wine buyers in 25 states – more than 70% of the United States market for off-premise premium wine sales. With its eight current logistic centers and the imminent opening of two additional centers, eVineyard will soon sell premium wines to each of the country's top 10 wine markets, reaching over 75% of the market and serving 27 states. With the recent acquisition of WineBuyer.com [http://www.WineBuyer.com] and its related technology, eVineyard is positioned for expansion in the B2B space. In addition, eVineyard will soon offer wine accessories in all 50 states through its recently announced marketing and fulfillment alliance with The Wine Appreciation Guild.

eVineyard offers a selection of over 5,000 domestic and imported premium wines and has exceptional features such as online education, promotions and loyalty programs, sweepstakes events, a wine club and free shipping with purchases over $25. eVineyard's superior selection and delivery capabilities have led to growth averaging more than 150% per quarter since launch. The company expects to achieve profitability in 2001.

17 November 2000

American restaurant site raises $42 million

An online reservation package for restaurant booking has raised $42 million to dispel doubts that the e-commerce sector might be having second thoughts.

OpenTable.com [http://www.OpenTable.com] which has just 1000 restaurants connected, plans to extend its offering of networked applications for the hospitality industry; delivering a full suite of Internet and software solutions to connect restaurants, hotels and consumers. OpenTable's technology is designed to help restaurants drive revenue and profitability, simplify operations, enhance guest relationships and effectively compete in today's marketplace. Investors in the company's financing round include leading venture capital firms Benchmark Capital, Adam Dell's Impact Venture Partners, Integral Capital, Epoch Partners, Comdisco Ventures and Upstart Capital.

A strategic investment comes from Worldspan – providers of global electronic distribution of travel information, Internet products and connectivity, and electronic commerce capabilities for travel agencies, travel service providers and corporations worldwide.

OpenTable will soon be releasing OTRestaurant – a web-based solution that enables restaurateurs to remotely access reservation data and reports, and dynamically update Dining Rewards in order to effectively and quickly manage yields.

Other OpenTable products currently in development include centralized database solutions to enable restaurant groups to efficiently manage reservations and customer databases for multiple locations, executive-level reporting applications, and Internet marketing tools.

17 November 2000

Safety police organise Net sweep to track down bogus animal feed

The US Department of Agriculture conducted a Surf Day to track down rogue online traders in animal feed.

They found nine companies selling 20 products. All of the companies have products for which they are making outright or implied drug claims that may not have been approved by the Food and Drug Administration. Seven are not licensed to do business in Washington. The companies identified by Department of Agriculture are based in Minnesota, Nevada, Oklahoma, North Dakota, Arizona, California, Australia and Germany.

17 November 2000

New York restaurants get their own B2B site

Restauranttrade.com launched in New York promising a full suite of electronic services.

RestaurantTrade.com [http://www.RestaurantTrade.com] manages order fulfillment and inventory, reporting and integration services. FoodTRENDS 2000, a study by American Express and Thomas Food Industry Register in America, projects that almost 90 percent of restaurants with fewer than six units will be going online to source and purchase food products and supplies by 2001. Further evidence of this trend is the fact that as of October 2000, RestaurantTrade.com has already signed up 10 independent restaurants and 25 suppliers in the New York market.

'Razor-thin margins have always been a problem for independent restaurants. We knew that there had to be a way to liberate chefs and managers from tedious paper pushing and enable them to focus on what they do best, which is create exceptional food and provide great service to their guests. With RestaurantTrade.com, we have done just that,' said Damian J. Mogavero, CEO and Founder.

The company's complete technology solution includes:

- Ordering and Receiving
- System-generated Faxing
- Price Requests
- Supplier Price Quotes
- Bin Management
- Inventory Management
- Supplier Catalog Maintenance
- Order History and Reporting
- User Management
- Online Customer Support
- Classifieds

Future enhancements to the solution will include:

- Point-of-Sale Integration and Reporting
- Menu Costing

- Direct Marketing for Suppliers
- Broadcast Specials

Initially launching in New York, RestaurantTrade.com plans to expand to new markets beginning in 2001. Beta testers and current customers of RestaurantTrade.com include suppliers that cover an estimated 90 percent of the New York market, such as D'Artagnan, Riviera Produce, Sam Tell & Sons, and E&S Wholesale Meats; as well as leading independent restaurants such as Gramercy Tavern, Union Pacific, Savoy, Tribeca Grill, Calle Ocho, Red Cat, Rain East and Rain West, and 71 Clinton Fresh Food; and renowned culinary schools such as the French Culinary Institute.

'RestaurantTrade.com is a significant step for the independent restaurant industry. We now have resources and tools that – unlike large chains – we were not able to develop on our own,' said Tom Colicchio, co-owner and executive chef of Gramercy Tavern and a RestaurantTrade.com customer. 'RestaurantTrade.com has become one of my best business partners. The solution is easy to use, saving me an invaluable amount of time every day.'

17 November 2000

Online deliveries rise 30% thanks to the storms

Online grocery ordering came into its own when severe weather lashed Chicago and the Mid West, sending orders up by 30% . . . but suppliers still got through.

Both Webvan-Chicago and HomeGrocer-Dallas reported that couriers created makeshift sleds to trek the totes through snow-covered streets and sidewalks. To make sure they did not track the slushy snow into the customers' homes, couriers used disposable shoe covers to place over their boots prior to entering their homes.

18 December 2000

Part 7
Net Tricks and Resources

The most revolutionary part of the Net are sites dedicated to transforming everyday tasks. They get less publicity than the high profile consumer dot coms but as tools for business this sector is likely to have a more fundamental impact on trading than any other.

Online help for startups

A new service aimed at getting people up to speed with starting a business has launched on the Net.

Steering well clear of e-commerce itself, Startups.co.uk [http://www.startups.co.uk] has posted background service features on how to start a new business, write a business plan, implications for tax and employment on the Net.

David Lester, a chartered accountant, said: 'It's high time the UK had a good entrepreneurial resource centre. The service is free to readers and funded by advertising and charges for entries in the searchable directories.'

10 January 2000

Texas offers the idiot cure

A Texas site offers cut-price tune-ups for websites and a full search engine registration program from an amazing $19.95.

One of the cardinal sins of the first generation of Web designers was omitting to put the tags in a website which were needed for a search engine to find and log the site. A Texas-based service Jjwebwonders.com [http://www.jjwebwonders.com] has the quick, cut-price remedy. It offers a tune up on the site and then a full registration programme.

It is only a four-man operation but has put on 240 companies in the last year out of its LaPorte offices. It offers four main services:

- Get Registered – submissions to 900 search engines for $19.95.
- A full tune up which includes registration but also revises the site and up dates the meta tags for $74.95.
- An ongoing programme is costed at $200
- and a reporting service on how high up the search engine your company name appears for a one off $29.

11 January 2000

Domain name pain

The cost of registering a domain name is about to come down.

Demon, the ISP provider, has slashed registration costs of a dedicated domain name from what it claims was £300, down to £50 until the end of February.

Moves across the industry look set to rationalise what so far has been a fluctuating market. While the Sunday Times classified adverts now offer to sell domain names like car number plates, behind-the-scenes deals look set to clean up the domain name scams. 'Dot com' names are universally charged at $70, an accepted American standard. The cost of other tags such as '.co.uk' or '.org' fluctuate according to who is charging.

Freenet.co.uk are giving them away free but you have to register with them. A small company in Norwich, ThinkTank.uk.net [http://www.thinktank.uk.net] is offering a combined service of a domain number plus an elementary web site for under £70. The service is initially aimed at the gift market and those wanting to make sure fathers can pass on the family domain as an heirloom to sons and daughters for birthdays but it can work just as effectively for a company logo.

11 January 2000

Now you can talk and see down the Net . . . for the cost of a local call

A new piece of Web brilliance allows surfers to talk to each other across the Net and if they have video cameras, even to see each other.

The free service is launched by Visitalk.com [http://www.visitalk.com] and can be downloaded from the Net for PCs – Macs will come later. Gizmos such as microphones and cameras are extras but the revolutionary new service will slash the price of video conferencing for small companies. Most interesting is that the cost of the call is charged only at local rate because it uses the Net rather than a conventional phone line, making international calls much cheaper.

19 January 2000

New gizmo reassures e-shoppers

Credit card insurers CPP have stepped into the e-shopping arena with a new scheme to cover online fraud from a credit card.

Currently most credit card companies expect customers to bear the first £50 of losses on a credit card fraud, and as most Internet shopping usually involves small amounts of money, then the real risk of losing out to a credit card scam can add up to about £100 a purchase. But CPP – the card protection company that allows terrestrial shoppers to notify any lost or stolen cards – has come up with a scheme called e-shopsafe [http:www.shopsafe.co.uk].

For a yearly subscription of £10, shoppers can cover their risk for as many as six credit cards against Internet fraud anywhere in the world. Current research suggests around 42% of Net users are concerned about using their cards on the Internet. There is only one loophole in the scheme: shoppers have to sign up online, which means their credit card details still have to be passed on to a Web browser. . .

02 May 2000

Psst ... wanna buy a second hand PC?

PC giant Dell has launched a new auction site to recycle second-hand computers.

Most of the stock [http://www.dellauction.co.uk] is in fact ex-leased computers that have been returned to the manufacturer and, according to Dell, have been fully refurbished. The headline catching bargain of the site is a Pentium III for a starting price of £1.

The official Dell models are backed up by service and support options, but the company is also promising soon a facility allowing anyone to sell on their second-hand computer equipment, and an email service to alert you when the site features a piece of kit that you have registered on your wishlist.

02 May 2000

Knock knock ... here comes the e-bailiff

Electronic debt collecting to tackle £4 billion small business overdraft.

Getaresult.co.uk [http://www.getaresult.co.uk/] is founded by the Credit Protection Association. The automated service sends out a series of three letters to follow up outstanding debts. The first warns that the CPA has been informed, the second is a reminder and the third threatens legal action. The CPA claims that a quarter of the 40,000 business collapses in the UK are due to late payments from debtors.

07 June 2000

Mind out for this new piece of red tape

EU directive changes the rules for e-selling.

A little-noticed piece of legislation known as the *Distance Selling Directive* slipped into operation on June 4. This sets out the terms and conditions for selling between parties that never see each other, including such media as the telephone, as well as the Net. The key parts cover the level of information given, legal confirmation of orders, the customer's right to withdraw within seven days, and legislating that the transaction should be completed within a month.

For websites, this means that they must display the following:

- name and address of supplier
- a description of the goods to be sold
- price and delivery costs
- arrangements for payment
- delivery and performance
- the right of withdrawal
- the price of the purchase including time spent on the Net
- how long the prices are valid and
- how long the contract lasts.

07 June 2000

Whoops! Is that your dotcom name for sale?

Huge auction of unconfirmed names threatens to undermine the market.

More than 12 million dotcom names are to be auctioned off in Virginia from companies that have either lapsed their subscriptions or not confirmed their ownership of names. Many UK companies could find themselves competing to buy back their own names from Internet speculators looking to cash in on the bonanza. Most names are likely to be more than five years old and may have been registered through companies that are no longer trading.

The bad news is that companies have only been given eight days to notify their concerns and interests...

04 July 2000

Interview with the Law: so who owns the website?

Top e-commerce lawyer launches legal backgrounders for worried companies and would be entrepreneurs.

The pitfalls and copyright issues that look to keep the legal trade rich for many years to come are predicted in a new newsletter from top e-commerce lawyer

Richard Boardman. For example writing in his latest issue Boardman concentrates on ownership. This is a sample extract to scare the life out of any company setting up its website for the first time. To sign up for more, email richardboardman@hotpop.com [mailto:richardboardman@hotpop.com].

'Once, in the heady days of the internet's commercial youth (about 3 or 4 years ago), a website build needed no more than an HTML editor and someone who could upload it to a bit of, probably paid-for, web space. Now though, e-commerce sites can cost up to half a million pounds to build and incorporate some sophisticated software, often written specially for that site . . . everything from animated graphics, to powerful database search engines through to intelligent knowledge management systems.

Suddenly, who owns what comes from that investment is an issue and how to manage and maintain the site matters as market-driven agility is no longer about a few HTML tag changes!!

The question is who should own what and does it really matter?

Well. . . ownership depends partly on commercial leverage, partly on need and partly on how smart and experienced each side's advisors are. Does it matter? . . . absolutely. For both website owners and developers the reasons why will soon become clear.

In a classic big build of a heavy e-commerce site you will often find a whole set of different IPRs. These include:

- Copyright, design right (and possibly database rights) in the look and feel . . . story boards, layout art work, site structure, navigation etc . . .
- Copyright in the basic HTML coding
- Copyright in the graphics files created by the developer
- Copyright in the graphics files created or supplied by the site owner
- Copyright in bought-in graphics
- Copyright in the software written by the developer including that to create interfaces with third party software (like Oracle), other programmers to make chat rooms and bulletin boards work etc.
- Copyright in the server-side applications handling such things as payments, data capture etc . . .
- Patentable ideas around site functionality

. . . and the list goes on depending on the particular site and the country involved. In some countries moral rights, personality rights and privacy rights will all be issues too.

As to who owns what: well, parking negotiability for a moment, at the very least the site owner has to own the look and feel to protect its investment. The HTML code generating that look and feel is more debateable. Great to own if you are the *site owner* because it makes it easier to stop copy sites, but also useful to own as the *developer* because it allows standard layouts to be re-used on other builds. The site owner should get the graphics files done for the site and the benefit of licences for third party graphics bought-in plus its own graphics provided for the build. Obviously Oracle and the rest get to keep the server-side applications, with the site owner taking out end-used licences and support (watch out for hidden costs here!).

Patentable ideas should be the site owner's too but if generated by the 'Generation x' development team, they may want to keep those tricks for sale to others and license them back.

The real problem is bespoke programmes written for the site but capable of resale without touching the look and feel. Is the site owner paying to own all the rights in those programmes or just access to them? Probably not but careful thought needs to go into commercial risk if these rights are handed over/not handed over. For the site owner, what about commercial risk, first-adopter advantage, return on development risk, access to source code for evolution and problem fixing? For the developer, what about lockout from future development tools, uncompetitive costing on ground-up new builds and the loss of the support lock-in advantage?

There is no perfect answer to this except to say it is well worth a good bit of negotiation on both sides and, if you are the site owner and you don't get the copyright, just a licence, then make sure you get some sector exclusivity and a decent support agreement with price movement restrictions, escrow for any non-open source and adaptation development rights for personal use, so you are not back asking for help to adapt the site in 6 months only to find there isn't anyone else on the tender list and guess what? . . . it's expensive!!

Finally, standard terms and conditions rarely work for large bespoke site builds, so avoid them on both sides. If the project is six figures up or high risk/value either way, talk detail!

04 July 2000

The true value of email lists

Yahoo coup opens the door on a whole new grove of olives.

Yahoo's acquisition of egroups.com [http://www.egroups.com] for a reputed $250 million raises interesting questions about Internet valuations. The email database of specialist subscribers is said to be the main value in the company as far as Yahoo is concerned.

Egroups, in the food area, have been pernicious exponents of what is effectively email advertising. This month's issue offers Greetings from the California Olive Industry. Get your Olive Brochure here! Under the UK Data Protection Act such messages would be illegal unless subscribers had specifically asked to be bombarded with such advertising messages. Laws in America are looser. Already in this country Sainsbury has flirted with the tactic.

The other question raised is how much real value there is in the supposedly stupendous list of people receiving such information. Efoodnews.com was surprised to find when it tried to register, that it was already registered – two years ago – and hadn't been back since. In fact, we had been trying to unsubscribe ever since without success, so perhaps Yahoo's supposed $25 valuation per email subscriber is a bit inflated in our case.

20 July 2000

Now beenz links down into the packaging

beenz.com has developed software especially for the food industry so that customers can earn points via the packaging and reclaim them on the Net.

'beenz can now bring to offline businesses the marketing benefits that we have brought to online businesses with integrated on- and offline incentive and reward programs and promotions. It's beenz everywhere,' said Charles Cohen, chairman and CEO of beenz.com Inc. '68% of consumers who use websites to research products later make their purchase in an offline store, and spend 30% more than other consumers do. beenzCodes will enable our customers to capture, track and tap into the spending power of the multi-channel consumer.'

beenzCodes are simple-to-enter codes issued inside product packaging by manufacturers or given to the purchaser at point of sale. They may also be used in direct mail to increase response rates or in print media to track performance. Each code is unique and allows the merchant to track the product from its point of manufacture to its purchase by the ultimate consumer.

The purchaser enters a code at a specified website, along with personal information that might be requested by the manufacturer, to earn beenz. The beenz earned by the consumer can be spent on products and/or services offered by the manufacturer, in online stores worldwide that form part of the beenz network, or transferred to the consumer's beenz rewardzcard MasterCard(tm) to spend anywhere that accepts MasterCard.

The company's launch of beenzCodes coincides with the acquisition by beenz.com of several key US and foreign patents. The patents, including patent #6,039,244 in the US and several pending in countries around the world, provide beenz with the exclusive right to place alphanumeric codes on or inside product packaging (hidden from the consumer before the purchase has been made), in conjunction with the purchaser earning beenz by submitting that code through a website. Other applications include retail and wireless devices.

beenz.com is headquartered in New York with offices across North America, Europe and the Asia-Pacific region. More information: www.beenz.com [http://www.beenz.com] or write to beenzcodes@beenz.com.

23 October 2000

Surprise Net link-up between Sainsbury and Oddbins

Sainsbury is to link-up with top-rated specialist off-licence Oddbins in a joint sales drive on the Net next year.

The move will take many industry watchers by surprise. Three thousand wines are planned to be sold via the Net, mail order and in-store as a joint venture.

01 December 2000

Index

4appetite.com, 189

AbestKitchen.com, 136
AboutFood.co.uk, 96
Advertising, 21, 113
Agribuys, 143, 146
AgriFist.com, 103
Agway, 5
AgWeb.com, 184
Allan, Chris, 8
Applebaum, Kevin, 179
Asda, 58, 69, 119, 164
Auctions, 118, 122

B2Bgalaxy.com, 145
Badger, Adam, 48
BagsOfTime.com, 54, 58
Barbieux, Yves, 131
Bascombe, Cheryl, 151
Banks, 13, 31
Banners, 31
Beans, 118
Beenz.com, 202
Beer, 7
BeerRitz.co.uk, 7
Berger, Roland, 106, 108
BestofBritish.com, 171
Bo.com, 36
Bovis, Jill, 17
Bradley, Carol, 155
Brake Brothers, 103
Brands, 5, 155, 174
Brier, Timothy, 185
Brivo, 125
BT, 171
BulkMarkets.com, 133
Button, Frankie, 46
BuyProduce.com, 85, 113

Cafés, 49
Calbier.com, 5
Campbell soups, 52
Candy.com, 138
Carrefour, 149
Cavallini, Stefano, 1
Celestial Seasonings, 64
CharlotteStreet.com, 23
ChateauOnline.com, 9
Chicken, 48
China, 121, 145
Chocolate, 25, 34, 178
Chocolate.co.uk, 25
Chopstix.net, 33
Claar, Curtis, 18
ClickonWine.com, 34
Coffee, 18, 19, 28, 29, 48, 54, 171, 176
CoffeeKid.com, 19
Coinstar, 67
Consumers Association, 157
Cooking, 94
Cookshop, 44
CooksIllustrated.com, 95
Corningsware-eschool.com, 95
Counsell, Richard, 3
Coupons.com, 148
CPGmarket.com, 105, 130
Credit cards, 13, 98, 105
Crites, Ken, 66
Cyber attacks, 33, 36, 50, 75, 76, 98, 105, 158
CyberCrop.com, 103

DairyBusiness.com, 147
Danone, 105, 130
Davis, Peter, 157
DellAuction.co.uk, 198
Del Monte, 68
Descartes, 169
DetConnect.com, 29
DHL, 28

Diet, 26, 29, 50, 63
DirectAg.com, 182
Directors, Institute of, 13
Domino Pizza, 8, 24
DropZone.co.uk, 166

Ecandy.com, 138
EcFood.com, 118, 122, 135
Efdex.com, 122, 156
Efoofmanageer.com, 91, 104, 108, 125
EggsBenefit.co.uk, 9, 43, 69
E-Insight, 51
Ellis, Anthony, 150
Email, 202
EmergeInteractive.com, 103
Engage Media, 147
Ernst & Young, 100
Espresso.com, 18
EthnicGrocer.com, 123
Europe, 101, 106 108
Euros, 41
Evineyard.com, 129, 190
Express Dairies, 67, 170

Farming, 2, 5, 14, 16, 17, 25, 73, 91, 103, 133, 147, 182, 192
FarmConnection.co.uk, 59
FarmingOnline.co.uk, 103
FarmPartner.com, 91
Farms.com, 182
FarmSource.com, 73, 182
Farrington, Paul, 26
Fish, 21, 47, 108, 131, 151
Fletcher Research, 112
FlyingNoodle.com, 10
Food.com, 104, 187
FoodandDrink.co.uk, 12
FoodBuy.com, 187
Food Distributors International, 107
FoodFerry.co.uk, 61
Food from Britain, 23, 69
Foodgalaxy.com, 145
FoodMarketExchange.com, 141
Food Marketing Institute, 107
FoodMarketPlace.com, 128
Foodndrink.co.uk, 12
FoodNet100.com, 121
Foodoo.com, 33, 35, 67
FoodSamples.com, 128
FoodTrader.com, 132
FoodTv.com, 2
FoodUSA.com, 179
FoodVision.com, 96, 136
Forrester, research, 36, 37, 101, 107, 181
Fraud, 98, 172, 197
FreeMarkets.com, 147
FreshdelMonte.com, 68
FreshFood.co.uk, 45, 72
FreshfromtheWeb.com, 148
Fruit, 85
Fruits.com, 68

Gate2Plate.com, 137
Gayton, Chris, 24, 33
GetaResult.co.uk, 198
GlobalNetExchange.com, 149
GMAbrands.com, 124
GM foods, 75
Gnx.com, 149
Godwin, Nick, 96
GoFish.com, 131, 151
Gomez, 61
GongShee.com, 123
Goodson, Talley, 147
Gourmet, 15, 33, 35, 47
Gourmet-Connect.com, 15
GourmetMarket.com, 188
Greenstuff.org.uk, 133

HandBag.com, 23
Hazlewood, 154
HealFarm.co.uk, 16
Heineken, 65
Herbs, 11
Hewitt, Patricia, 23
Home delivery, 8, 10, 54, 58, 74, 108, 116, 180, 194
HomeChocolateFactory.com, 34
HomeGrocer.com, 116, 178, 194
HomePort, 125
Hotels, 8, 10

Iceland, 63, 159
Imeson, Robert, 56
InfusionsCoffee.co.uk, 48
Intelligent software, 111

Interviews, 3, 19, 59, 82, 83, 96, 164, 167, 184
Ismail, Abdul, 17
Ivillage.co.uk, 42

Jackson, Michael, 56
Jankowski, Geoff, 125
JekkasHerbFarm.co.uk, 11
Jet, 166
Jobs, 16
Just-drinks.com, 51
Just-food.com, 51

Kellmerit, Daniel, 92
Kenyon, Peter, 1
Kirkham, Peter, 15
Kite mark, 23
Knight, Laura, 40
KooBuyCity.co.uk, 54, 58
Korea, 120
Kozmo.com, 58, 176, 180
Kraft, 123

Lantes, 125
LastMileSolutions, 60, 161, 169
LastMinute.com, 10
LastOrders.com, 43
LateRooms.com, 8, 10
Laughlin, Joe, 149
Law, 199
LeapingSalmon.com, 1
Learning, 5
LetsBuyIt.com, 36
Lewis, Seth, 9, 11
Londis, 166
LowCarbTreats.com, 27

MadAboutWine.com, 117
Majestic, 29
Malamatines, Denis, 39
Marks & Spencer, 160
Marshall, James, 1
Martin, David, 164
Martin, Jeff, 21
Martins-Seafresh.co.uk, 47
Mathews, Bernard, 68
M-box, 170
Meals.com, 66, 148
Megson, Angela, 6
Melin, John, 148
MexGrocer.com, 65
Mogavero, Damian, 193
Monsanto, 73

McCormick spices, 62
McDonalds, 104, 117, 120, 151
McKown, Steve, 64
McVicar, Jekka, 11

Namaste.com, 123
Names, 12, 196, 199
Neilsen AC, 2
Nestle, 50, 66, 105, 130, 147, 148
Newton, Malc, 83
Noodles, 10
NovoPoint.com, 144
NutriSystem.com, 50, 64
Nyt.com, 95

Oliver, Jamie, 29
O'Mullane, Fergal, 110
OpenTable.com, 191
Organic, 3, 45, 46, 56, 72, 74, 110, 132
Organic Consumers Association, 75
OrganicFarmFoods.co.uk, 74
Orgasmic wine, 22

Palm, 169
Parsons, Chuck, 18
Payroll, 117
PDQuick.com, 4
Perrin, Marc, 11, 14
Petch, Ann, 16
Phones, 13, 27, 32
Pink Dot, 4, 82
Pizza, 8, 24
PlanetRice.net, 140
Point of Sale Ltd, 143
Pordum, Keith, 30
Porter, Nicola, 25
PoultryScotland.co.uk, 17
PR, 47, 52, 82
Pratt, Bill, 167
Priceline.com, 10, 39, 185
Prince, Mark, 19
Pringles, 5

Pubs, 42, 65
PureFood.org, 75

QueRico.com, 123
QueueJumper.com, 180

RealBeer.com, 96
Recipeweb.co.uk, 38
Restaurants, 9, 12, 43, 183, 186, 187, 190, 191, 193
RestaurantTrade.com, 183
Rice, 140
RichClickings.com, 23
RoadNet.com, 183
Rooster.com, 182

Safeway, 50, 162, 166, 169
Sainsbury, 5, 61, 115, 149, 150, 156, 157, 160, 166, 203
Sara Lee, 139
ScottSargeant.com, 40
Seals, Mark, 67
SellMeat.com, 182
ShopSmart.com, 119
SimplyOrganic.net, 56, 72
Somerfield, 170
SomersetOrganics.co.uk, 3
Spain, 24
SpanishFoodDirect.com, 24
SpecialityFoods.co.uk, 23
Spices, 17, 62
Sprieser, Judy, 129
StartUps.co.uk, 195
Stenger, Jan, 92, 104, 109
Streamline.com, 189
SurplusFood.com, 128
Sweetener, 122
SweetMart.co.uk, 17
SynergyDiet.com, 26

Taste.co.uk, 150
Tavolo.com, 95, 179
Tea, 65
Telefonica, 145
Tesco, 38, 42, 51, 61, 66, 71, 108, 142, 153, 154, 156, 158, 159, 160, 161, 163, 164
Thailand, 141, 147, 148
Tomasino, Frank, 132
TradingProduce.com, 93, 120, 127
Transora.com, 129
Turkey, 68

Unilever, 16
UrbanFetch.co.uk, 54, 58, 180
Users, statistics, 32, 37, 44, 47, 51, 52, 55, 100, 101, 105-7, 173-4

Valhrona, 25
Veg-Net.co.uk, 25
Verdict, 52, 105
Vintners.com, 140
Virgin, 22, 36, 40
VirginWines.com, 22
VirtualGourmet.com, 188

Waitrose, 60, 71, 74, 167
Walmart, 54, 58, 119
Webcast, 139
WebVan, 66, 116, 175, 194
Whittards, 28, 171
Wine, 9, 11, 14, 18, 22, 29, 34, 40, 57, 129, 133, 134, 139, 171, 188, 190, 203
WineandCo.com, 9, 11, 14, 40, 57, 117
WineBid.com, 18
WineBuyer.com, 129
WorldofFruit.com, 110, 114
WorldJuiceTrade.com, 134
WorldwineTrade.com, 133

Xsag.com, 182
Young, Thoby, 45

Zbox, 125
ZeroCarb.com, 27